A Level
Chemistry
for OCR
Year 2
Revision Guide

B

David Goodfellow
Mark Gale

OXFORD
UNIVERSITY PRESS

OXFORD
UNIVERSITY PRESS

Great Clarendon Street, Oxford, OX2 6DP, United Kingdom

Oxford University Press is a department of the University of Oxford. It furthers the University's objective of excellence in research, scholarship, and education by publishing worldwide. Oxford is a registered trade mark of Oxford University Press in the UK and in certain other countries

© David Goodfellow and Mark Gale

The moral rights of the authors have been asserted

First published in 2017

British Library Cataloguing in Publication Data
Data available

978-0-19-835779-7

10 9 8 7 6 5 4 3 2 1

Paper used in the production of this book is a natural, recyclable product made from wood grown in sustainable forests. The manufacturing process conforms to the environmental regulations of the country of origin.

Printed in Great Britain

Acknowledgements

Cover: EYE OF SCIENCE/SCIENCE PHOTO LIBRARY

Artwork by Q2A Media

Although we have made every effort to trace and contact all copyright holders before publication this has not been possible in all cases. If notified, the publisher will rectify any errors or omissions at the earliest opportunity.

f Energy needed to break one C–Cl bond =
$\frac{340000}{6.02 \times 10^{23}} = 5.64 \times 10^{-19} \, J$ [1]

Minimum frequency required $= \frac{E}{h} =$

$\frac{5.64 \times 10^{-19}}{6.63 \times 10^{-34}} = 8.52 \times 10^{14} \, Hz$ [1]

Frequency of shortest wavelength of UVA $= \frac{c}{\lambda} =$

$\frac{3.0 \times 10^{8}}{315 \times 10^{-9}} = 9.52 \times 10^{14} \, Hz$ [1]

Frequency of longest wavelength of UVA =

$\frac{3.0 \times 10^{8}}{400 \times 10^{-9}} = 7.50 \times 10^{14} \, Hz$ [1]

So only part of the UVA range is able to cause the C–Cl bond to break [1]

5 a i 2-hydroxypropanoic acid [1]

ii ethanol [1] conc sulfuric acid AND reflux [1]

b i distill off at 155°C [1]

ii any <u>anhydrous</u> salt e.g. anhydrous sodium sulfate [1]

c New product will not have a (broad) peak at 2500–3300 cm^{-1} [1] because O–H has reacted (to form ester group) [1] Peak at 1700–1725 cm^{-1} (due to C=O in carboxylic acid) will now be at 1735–1750 (due to C=O in ester) [1]

d *The following analysis of data could be expected:*

Mass spectrum:
M_r of empirical formula unit = 162
So molecular formula = empirical formula

Infrared: Broad peak between 3200 and 3600 indicates OH in alcohol 1790 indicates CO in acid anhydride? No very broad peak 2500–3200 / no peak between 1720-1725 indicates no COOH

NMR

3 carbon environments
δ = 220 due to C=O, δ = 85 due to C–OH, δ = 35 due to CH$_3$–C
3 proton environments
δ = 1.8 = CH$_3$–R
δ = 4.2 = CH–O
δ = 11.5 therefore is OH

5–6 marks:

Correctly deduces structure
Analyses all types of data (mass spectrum, infrared spectrum, nmr)
Explains in detail how each feature of the molecule relates to the data 3–4 marks:

Correctly deduces most feature of the structure
Analyses most types of data
Explains how most features of the molecule relate to the data 1–2 marks:

Deduces some features of the molecule
Analyses some of the data *Explains how some features of the molecule relate to the data*

e i

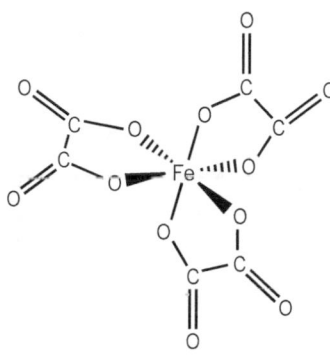

[1] condensation polymerisation AND a small molecule / water is lost [1]

ii lactic acid can be obtained from renewable natural sources AND raw materials for nylon come from crude oil [1]

f i $Ca(OH)_2 + H_2C_2O_4 \rightarrow Ca(C_2O_4) + 2H_2O$ [1]

ii

Correct structure (see above) [1] Possesses two lone pairs of electrons [1] Which it can donate / use to form bonds to one metal ion [1]

iii $[Fe(C_2O_4)_3]^{3-}$ formula [1] charge [1]

iv Structure is:

Octahedral arrangement of bonds around central Fe (with correct 3-dimensional representation) [1] Three C$_2$O$_4$ groups attached in a geometrically feasible arrangement [1]

g i ethanedioic acid system has a more negative electrode potential than manganate system [1]

ethanedioic acid donates electrons / is oxidised [1]

manganate accepts electrons / is reduced [1]

ii purple colour disappears [1] bubbling [1]

iii moles of manganate $= 19.1 \times \frac{0.02}{1000} = 3.82 \times 10^{-4}$ [1] moles of $H_2C_2O_4 = 9.55 \times 10$ [1]

mass of $H_2C_2O_4$ in 250 cm^3 = 9.55 10^{-3} × 90 = 0.8595 g [1] % ethanedioic acid in sample $= 0.8595 \times \frac{100}{0.98} = 87.7$ [1] answer to 2 s.f. = 88 [1]

Periodic table

Key
atomic number
Symbol
name
relative atomic mass

(1)	(2)											(3)	(4)	(5)	(6)	(7)	(0)
1																	**18**
1 **H** hydrogen 1.0	**2**											**13**	**14**	**15**	**16**	**17**	2 **He** helium 4.0
3 **Li** lithium 6.9	4 **Be** beryllium 9.0											5 **B** boron 10.8	6 **C** carbon 12.0	7 **N** nitrogen 14.0	8 **O** oxygen 16.0	9 **F** fluorine 19.0	10 **Ne** neon 20.2
11 **Na** sodium 23.0	12 **Mg** magnesium 24.3	**3**	**4**	**5**	**6**	**7**	**8**	**9**	**10**	**11**	**12**	13 **Al** aluminium 27.0	14 **Si** silicon 28.1	15 **P** phosphorus 31.0	16 **S** sulfur 32.1	17 **Cl** chlorine 35.5	18 **Ar** argon 39.9
19 **K** potassium 39.1	20 **Ca** calcium 40.1	21 **Sc** scandium 45.0	22 **Ti** titanium 47.9	23 **V** vanadium 50.9	24 **Cr** chromium 52.0	25 **Mn** manganese 54.9	26 **Fe** iron 55.8	27 **Co** cobalt 58.9	28 **Ni** nickel 58.7	29 **Cu** copper 63.5	30 **Zn** zinc 65.4	31 **Ga** gallium 69.7	32 **Ge** germanium 72.6	33 **As** arsenic 74.9	34 **Se** selenium 79.0	35 **Br** bromine 79.9	36 **Kr** krypton 83.8
37 **Rb** rubidium 85.5	38 **Sr** strontium 87.6	39 **Y** yttrium 88.9	40 **Zr** zirconium 91.2	41 **Nb** niobium 92.9	42 **Mo** molybdenum 95.9	43 **Tc** technetium	44 **Ru** ruthenium 101.1	45 **Rh** rhodium 102.9	46 **Pd** palladium 106.4	47 **Ag** silver 107.9	48 **Cd** cadmium 112.4	49 **In** indium 114.8	50 **Sn** tin 118.7	51 **Sb** antimony 121.8	52 **Te** tellurium 127.6	53 **I** iodine 126.9	54 **Xe** xenon 131.3
55 **Cs** caesium 132.9	56 **Ba** barium 137.3	57–71 lanthanoids	72 **Hf** hafnium 178.5	73 **Ta** tantalum 180.9	74 **W** tungsten 183.8	75 **Re** rhenium 186.2	76 **Os** osmium 190.2	77 **Ir** iridium 192.2	78 **Pt** platinum 195.1	79 **Au** gold 197.0	80 **Hg** mercury 200.6	81 **Tl** thallium 204.4	82 **Pb** lead 207.2	83 **Bi** bismuth 209.0	84 **Po** polonium	85 **At** astatine	86 **Rn** radon
87 **Fr** francium	88 **Ra** radium	89–103 actinoids	104 **Rf** rutherfordium	105 **Db** dubnium	106 **Sg** seaborgium	107 **Bh** bohrium	108 **Hs** hassium	109 **Mt** meitnerium	110 **Ds** darmstadtium	111 **Rg** roentgenium	112 **Cn** copernicium		114 **Fl** flerovium		116 **Lv** livermorium		

57 **La** lanthanum 138.9	58 **Ce** cerium 140.1	59 **Pr** praseodymium 140.9	60 **Nd** neodymium 144.2	61 **Pm** promethium 144.9	62 **Sm** samarium 150.4	63 **Eu** europium 152.0	64 **Gd** gadolinium 157.2	65 **Tb** terbium 158.9	66 **Dy** dysprosium 162.5	67 **Ho** holmium 164.9	68 **Er** erbium 167.3	69 **Tm** thulium 168.9	70 **Yb** ytterbium 173.0	71 **Lu** lutetium 175.0
89 **Ac** actinium	90 **Th** thorium 232.0	91 **Pa** protactinium	92 **U** uranium 238.1	93 **Np** neptunium	94 **Pu** plutonium	95 **Am** americium	96 **Cm** curium	97 **Bk** berkelium	98 **Cf** californium	99 **Es** einsteinium	100 **Fm** fermium	101 **Md** mendelevium	102 **No** nobelium	103 **Lr** lawrencium

AS/A Level course structure

This book has been written to support students studying for OCR A Level Chemistry B (Salters). It covers the Year 2 content from the specification. The content is arranged by chemical idea. The content covered is shown in the contents list, which also shows you the page numbers for the main topics within each chapter.

AS exam

Year 1 content

1 Elements of life
2 Developing fuels
3 Elements from the sea
4 The ozone story
5 What's in a medicine

Year 2 content

6 The chemical industry
7 Polymers and life
8 Oceans
9 Developing metals
10 Colour by design

A level exam

A Level exams will cover content from Year 1 and Year 2 and will be at a higher demand. You will also carry out practical activities throughout your course.

Contents by chemical idea

How to use this book		**v**

4 Energy changes and chemical reactions — 2
4.4   Ions in solution (O1) — 2
4.5   Enthalpy and entropy (O5) — 6
     Practice questions — 9

5 Structure and properties — 11
5.5   Protein structure (PL 5) — 11
5.6   Molecular recognition (PL 7) — 13
5.7   Bonding dyes to fibres (CD 6) — 15
     Practice questions — 17

6 Radiation and matter — 18
6.6   Further mass spectrometry (PL9) — 18
6.7   Nuclear magnetic resonance (NMR) spectroscopy (PL 9) — 20
6.8   Using combined spectroscopic techniques to deduce the structure of organic molecules (PL9) — 24
6.9   Coloured organic molecules (CD 1) — 26
     Practice questions — 28

7 Equilibrium in chemistry — 30
7.5   Equilibrium constant K_c, temperature, and pressure (CI 2) — 30
7.6   Solubility equilibria (O4) — 32
7.7   Gas–liquid chromatography (CD 8) — 33
     Experimental techniques — 35
     Practice questions — 36

8 Acids and bases — 37
8.2   Strong and weak acids and pH (O2) — 37
8.3   Buffer solutions (O3) — 40
     Experimental techniques — 42
     Practice questions — 43

9 Redox — 44
9.3   Redox reactions, cells, and electrode potentials (DM 4) — 44
9.4   Rusting and its prevention (DM 5) — 48
     Experimental techniques — 50
     Practice questions — 51

10 Rates of reactions — 52
10.5   Rates of reactions (CI 3) — 52
10.6   Rate equation of a reaction (CI 4) — 53
10.7   Finding the order of reaction with experiments (CI 5) — 55
10.8   Enzymes (PL 6) — 58
     Practice questions — 60

11 The periodic table — 61
11.4   The d-block (DM1, DM3, DM6) — 61
11.5   Nitrogen chemistry (CI1) — 64
     Experimental techniques — 66

12 Organic chemistry: frameworks — 67
12.4   Arenes (CD 2, CD3) — 67
12.5   Reactions of arenes (CD 4) — 70
12.6   Azo dyes (CD 5) — 73
     Practice questions — 75

13 Organic chemistry: modifiers — 76
13.5   Carboxylic acids (PL 1) — 76
13.6   Amines (PL 2) — 77
13.7   Hydrolysis of amides and esters (PL 3) — 79
13.8   Amino acids, peptides, and proteins (PL 4) — 80
13.9   Oils and fats (CD 7) — 83
13.10   Aldehydes and ketones (CD 9) — 84
     Experimental techniques — 86
     Practice questions — 88

14 Chemistry in industry — 89
14.3   Operation of a chemical manufacturing process (CI 6) — 89
     Practice questions — 92

15 Human impacts — 93
15.3   Radiation in, radiation out (O2) — 93
     Practice questions — 96

16 Molecules in living systems — 97
16.1   DNA and RNA (PL 8) — 97
     Practice questions — 101

17 Organic synthesis — 103
17.1   Functional group reactions (CD 10) — 103
17.2   Classification of organic reactions (CD 11) — 105
     Practice questions — 107

Answers to summary questions — 108
Answers to practice questions — 114
Synoptic questions — 117
Answers to synoptic questions — 121
Periodic table — 124

How to use this book

This book contains many different features. Each feature is designed to support and develop the skills you will need for your examinations, as well as foster and stimulate your interest in chemistry.

This book is structured by chemical idea.

 Worked example

Step-by-step worked solutions.

Common misconception

Common student misunderstandings clarified.

 Go further

Familiar concepts in an unfamiliar context.

Maths skill

A focus on maths skills.

Model answers

Sample answers to exam-style questions.

Summary Questions

1 These are short questions at the end of each topic.

2 They test your understanding of the topic and allow you to apply the knowledge and skills you have acquired.

3 The questions are ramped in order of difficulty. Lower-demand questions have a paler background, with the higher-demand questions having a darker background. Try to attempt every question you can, to help you achieve your best in the exams.

Specification references

→ At the beginning of each topic there is a specification reference to allow you to monitor your progress.

Key term

Pulls out key terms for quick reference.

Revision tips

Revision tips contain prompts to help you with your understanding and revision.

Synoptic link

These highlight the key areas where topics relate to each other. As you go through your course, knowing how to link different areas of chemistry together becomes increasingly important. Many exam questions, particularly at A Level, will require you to bring together your knowledge from different areas.

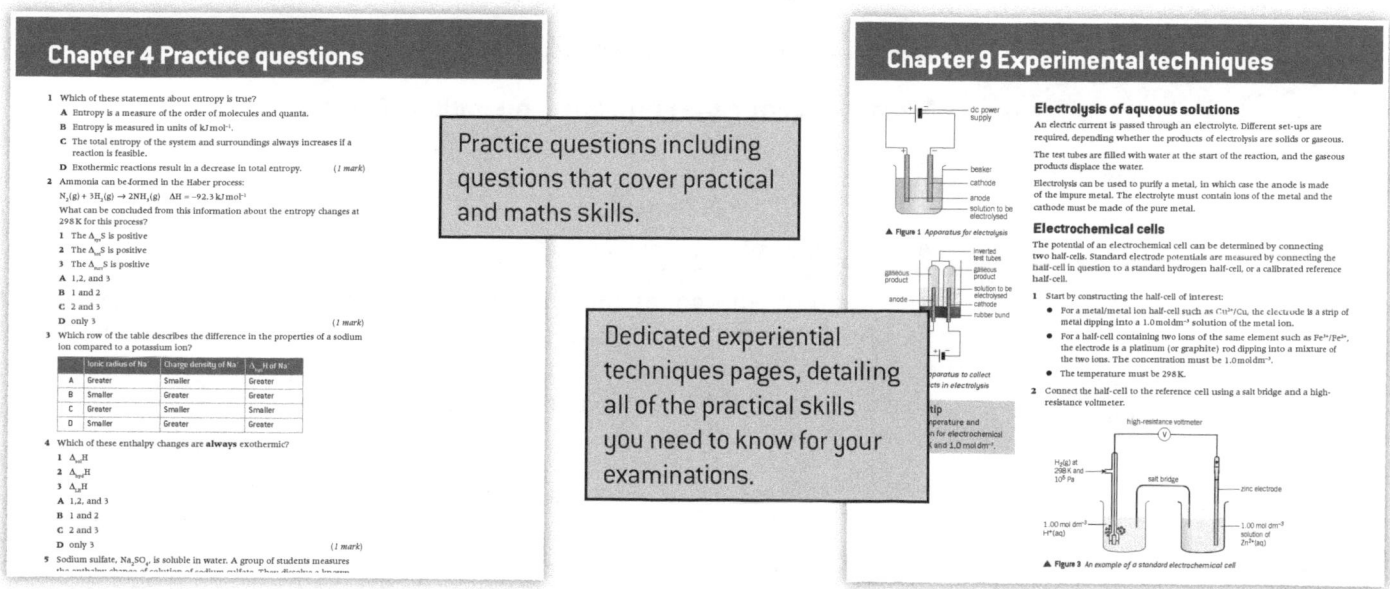

Practice questions including questions that cover practical and maths skills.

Dedicated experiential techniques pages, detailing all of the practical skills you need to know for your examinations.

4.4 Ions in solution

Revision tip

Notice that although you can imagine the dissolving process involves the breaking up of an ionic lattice – an endothermic process – the lattice enthalpy refers to the formation of an ionic lattice and is therefore **exothermic**.

Key terms

Enthalpy change of solution, $\Delta_{sol}H$: The enthalpy change when 1 mole of solute dissolves to form a solution, e.g.

$CaCl_2(s) + aq \rightarrow Ca^{2+}(aq) + 2Cl^-(aq)$

Lattice enthalpy, $\Delta_{LE}H$: The enthalpy change when 1 mole of an ionic solid is formed from gaseous ions, e.g.

$Ca^{2+}(g) + 2Cl^-(g) \rightarrow CaCl_2(s)$

Enthalpy change of hydration, $\Delta_{hyd}H$: The enthalpy change when 1 mole of a gaseous ion is hydrated by forming bonds to water molecules, e.g.

$Ca^{2+}(g) + aq \rightarrow Ca^{2+}(aq)$

Hydrated ion: An ion bonded to water molecules, e.g.

$Ca^{2+}(aq)$

▼ **Table 1** *Enthalpy of hydration values for some cations*

Ion	$\Delta_{hyd}H$/kJ mol^{-1}
Li$^+$	−520
Na$^+$	−406
K$^+$	−320
Rb$^+$	−296
Mg^{2+}	−1926
Ca^{2+}	−1579
Sr^{2+}	−1446
Al^{3+}	−4680

Dissolving ionic compounds

Many (but not all) ionic compounds are soluble in water. The ionic lattice breaks up and new bonds are formed between water molecules and the separate ions.

Enthalpy level diagrams are drawn to illustrate the energy changes involved in dissolving processes.

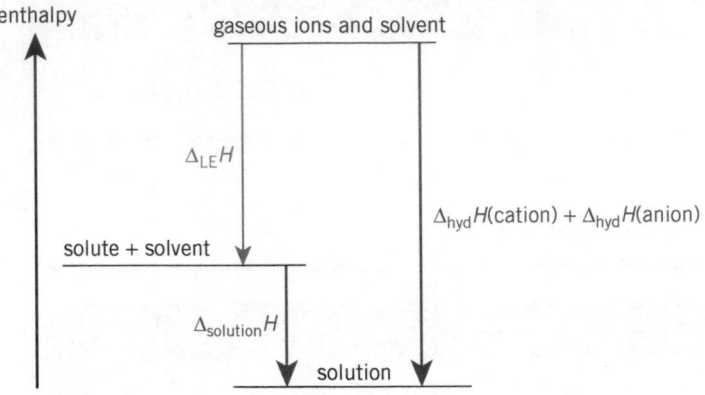

▲ **Figure 1** *An enthalpy level diagram for the dissolving of an ionic compound*

Enthalpy changes involved in dissolving

Figure 1 shows the relationship between the lattice enthalpy, the enthalpy change of solution, and the enthalpy changes of hydration of the positive and negative ions.

Hydrated ions

Hydrated ions are surrounded by water molecules that bond to them by ion–dipole bonds.

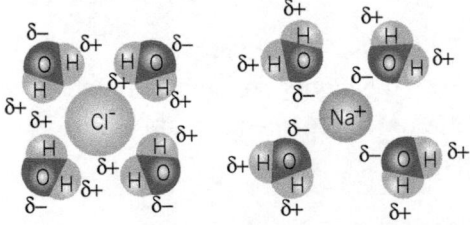

▲ **Figure 2** *Hydrated chloride and sodium ions*

Factors affecting the enthalpy of hydration of ions

Enthalpy of hydration will be more negative if

- the ionic charge is greater
- the ionic radius is smaller.

Alternatively, you can say that the enthalpy of hydration is more negative if the **charge density** of an ion is greater. Increased charge density means that an ion attracts more water molecules, and forms stronger ion–dipole bonds. So the energy released by bond forming is greater.

Factors affecting the lattice enthalpy

Lattice enthalpy is also affected by ionic charge and size of ions, following a similar pattern to enthalpy of hydration.

Lattice enthalpy becomes more negative if:

- the ions in the lattice are more highly charged
- the ions in the lattice are smaller.

So lattice energy is more negative if the ions in an ionic lattice have a greater charge density.

▼ **Table 2** *Enthalpy of lattice enthalpy values for some ionic compounds*

Compound	$\Delta_{LE}H$/kJ mol^{-1}	Compound	$\Delta_{LE}H$/kJ mol^{-1}
Li_2O	−2806	MgO	−3800
Na_2O	−2488	CaO	−3419
K_2O	−2245	SrO	−3222
LiF	−1047	MgF_2	−2961
NaF	−928	CaF_2	−2634
KF	−826	Al_2O_3	−15 916

Synoptic link

Charge density was introduced to explain the patterns in the thermal decomposition of carbonates in Topic 11.2, The s-block: Groups 1 and 2.

Synoptic link

Calculations involving these enthalpy changes may make use of Hess' Law. This was introduced in Topic 4.2, Enthalpy cycles.

Model answer: Calculating enthalpy changes involved in the dissolving process

An enthalpy level diagram for the dissolving of calcium hydroxide is shown in Figure 3.

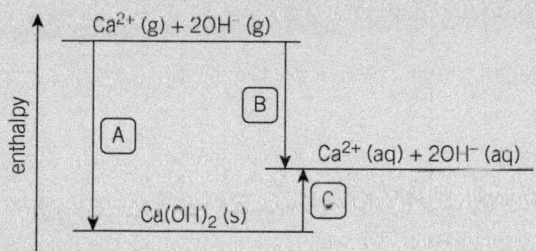

▲ **Figure 3** *Enthalpy cycle for the dissolving of calcium hydroxide*

Use the data below to calculate the enthalpy change of solution of calcium hydroxide, $\Delta_{sol}H$ [Ca(OH)$_2$]

$\Delta_{LE}H$[Ca(OH)$_2$] = −2506 kJ mol^{-1}

$\Delta_{hyd}H$ [Ca^{2+}(aq)] = −1579 kJ mol^{-1}

$\Delta_{hyd}H$ [OH$^-$(aq)] = −460 kJ mol^{-1}

1. Identify the data that tells you the value of enthalpy change A. This is the lattice enthalpy, so A = −2506 kJ mol^{-1}.
2. Identify the data that tells you the value of enthalpy change B. This is the sum of the enthalpy changes of hydration of 1Ca^{2+} + 2OH$^-$, so B = −1579 + 2 × −460 = −2499 kJ mol^{-1}
3. Use Hess's Law to write down the relationship between A, B, and C. A = B − C, so C (the enthalpy of solution) = B − A
4. Calculate C: C = −2499 − (−2506) = +7 kJ mol^{-1}

Make sure that you give the value a sign, and that the sign is consistent with the diagram, which shows an endothermic reaction.

Measuring enthalpy changes of solution experimentally

Lattice enthalpies and enthalpy changes of hydration cannot be measured directly in experiments. However enthalpy changes of solution can be measured using the method described in Chapter 4 Experimental techniques.

A known mass of solid is dissolved in a known mass of solvent and the temperature change is measured.

The energy exchanged with the water is calculated using the equation $E = mc\Delta T$.

This number is then scaled up to find the energy exchanged when 1 mole of solute dissolves.

The relationship between enthalpy change of solution and solubility

If the energy released by forming bonds in the hydration of ions can compensate for the energy required to break up the lattice, then an ionic solute will be soluble.

So enthalpy changes of solution are either exothermic or very slightly endothermic.

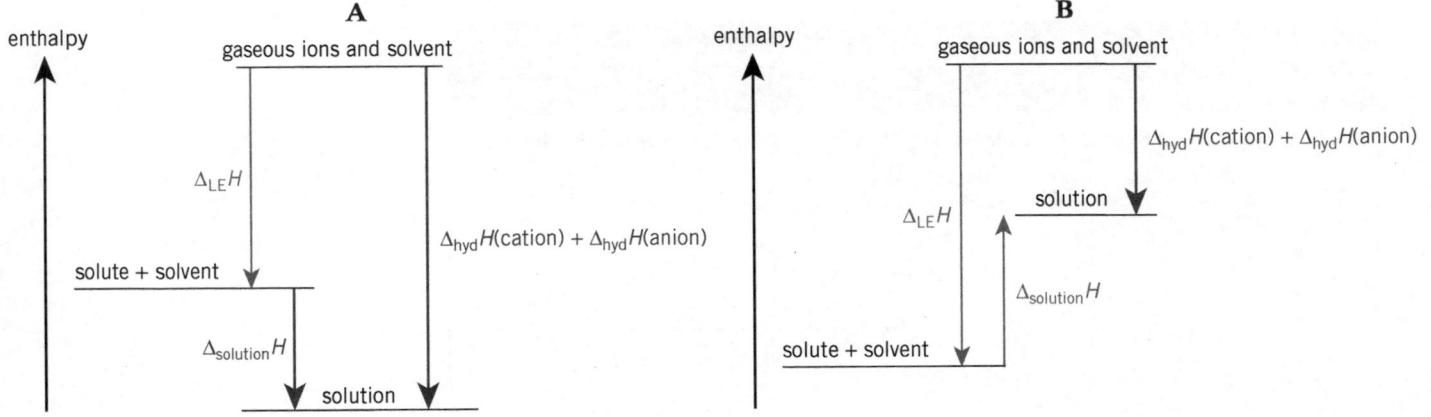

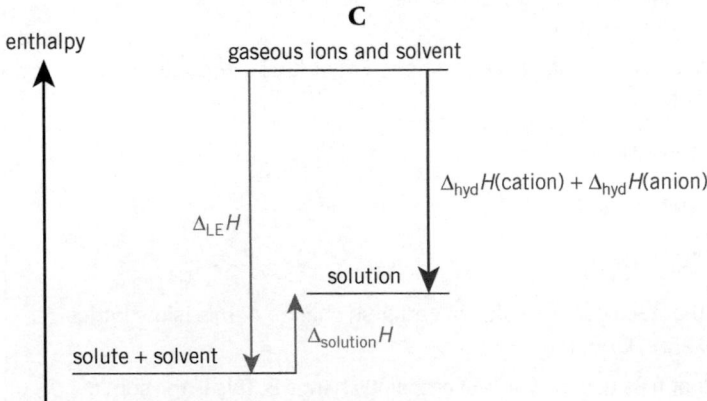

▲ **Figure 4** *Enthalpy cycles for exothermic and endothermic dissolving processes (soluble and insoluble). Cycle A (exothermic) is likely to be soluble, Cycle B (very endothermic) will be insoluble and Cycle C (slightly endothermic) may be soluble*

Model answer: Explaining the solubility of ionic substances

Many ionic substances are soluble in water. Explain why.

- Ionic bonds are broken when an ionic substance dissolves. Some hydrogen bonds also break between water molecules.

- Ion–dipole bonds form between water molecules and the free ions.

- The strength of the bonds formed is similar to the strength of the bonds broken.

- So the energy released by bond formation is sufficient to compensate for the energy required to the break the bond between ions.

> Describe the bonds broken during dissolving.

> Describe the bonds that can form when the substance dissolves.

> Compare the strength of the bonds broken and formed. Be careful to use precise language – common errors are to be imprecise and to say things like 'more energy is needed to make bonds than to break them' **OR** 'more bonds are broken than made'.

> Relate strength of bonds to energy changes.

You may also be asked to discuss why ionic substances do not dissolve in non-polar solvents such as cyclohexane. The difference here is that the ion–dipole bonds formed between the solvent molecules and the ions are much weaker than the bonds that are broken.

Model answer: Explaining why ionic compounds are insoluble in non-polar solvents

Sodium chloride does not dissolve in cyclohexane. Explain why.

- In order to dissolve, ionic bonds would need to break between the ions in the sodium chloride lattice. Instantaneous dipole–induced dipole bonds would also need to break between the cyclohexane molecules.

- Because cyclohexane molecules do not have a permanent dipole, only weak ion–dipole bonds could form between cyclohexane molecules and the free ions.

- The strength of the bonds that could form is much weaker than the bond that would need to break.

- So the energy released by bond formation is not sufficient to compensate for the energy required to the break the bond between ions.

> Imagine that the compound does dissolve and describe the bonds broken during dissolving.

> Describe the bonds that could form if the substance dissolves.

> Compare the strength of the bonds broken and formed.

> Relate strength of bonds to energy changes.

Summary questions

1 Magnesium chloride, $MgCl_2$, dissolves in water in an exothermic reaction. Draw out a diagram to show the energy changes involved in the dissolving process. Label the substances involved and the enthalpy changes.

(7 marks)

2 Calcium chloride is soluble in water but does not dissolve in the organic solvent cyclohexane. Explain why. *(4 marks)*

3 a Use the diagram you drew for question 1, and the data below, to calculate a value for the enthalpy of hydration per mole of chloride ions:

$\Delta_{LE}H[MgCl_2]^- = -2526 \, kJ \, mol^{-1}$

$\Delta_{hyd}H[Mg^{2+}(aq)] = -1926 \, kJ \, mol^{-1}$

$\Delta_{sol}H[MgCl_2(s)] = -155 \, kJ \, mol^{-1}$ *(4 marks)*

b The enthalpy of hydration of the Ca^{2+} ion is $-1579 \, kJ \, mol^{-1}$. Discuss why this value is less negative than that for the Mg^{2+} ion. *(2 marks)*

Revision tip

You can also use the same ideas to explain why two covalent liquids will mix. Simply list the type of intermolecular bonds broken and formed and compare the strength of these bonds.

Synoptic link

A complete explanation for solubility of solutes requires discussion of entropy changes. Entropy is discussed in Topic 4.5, Enthalpy and entropy.

4.5 Enthalpy and entropy

Specification reference: O (d), (e), (f), (g)

Entropy

Entropy of particles and feasible processes

A feasible process is one that will occur without any energy input – although it may occur very slowly.

Many feasible processes involve particles becoming more disordered – liquids or gases tend to mix and solids tend to dissolve.

Entropy is a measure of the disorder of these particles or of energy quanta. It is calculated by working out the number of ways of arranging the particles or energy quanta.

▼ **Table 1** *When water changes state, the entropy of the particles increases*

State		Entropy S / $J K^{-1} mol^{-1}$
increasing disorder	$H_2O(s)$	+41
	$H_2O(l)$	+70
	$H_2O(g)$	+189

Differences in the entropy of solids, liquids, and gases

Particles in a solid are rigidly fixed in place, whereas particles in a gas are free to move around and take up many different positions. This means that there are more ways of arranging the particles.

In general gases have a much higher molar entropy than liquids, which in turn have a higher molar entropy than solids.

Entropy changes

The total change in entropy associated with a process is given the symbol $\Delta_{tot}S$.

This is the sum of two entropy changes:

- The entropy change in the system, $\Delta_{sys}S$, which is due to the change in the number of ways of arranging particles.

- The entropy change in the surroundings , $\Delta_{surr}S$, which is due to the change in the number of ways of arranging energy quanta that are exchanged with the surroundings.

So $\Delta_{tot}S = \Delta_{sys}S + \Delta_{surr}S$

Calculating entropy

Calculating $\Delta_{sys}S$ and $\Delta_{surr}S$

$\Delta_{sys}S + \Delta_{surr}S$ can be calculated given appropriate data.

- $\Delta_{sys}S$ is calculated using values of molar entropy, S:

$$\Delta_{sys}S = \Sigma S \text{ (products)} - \Sigma S \text{ (reactants)}$$

- $\Delta_{surr}S$ varies with temperature, but can be calculated from the temperature in K and the enthalpy change for the reaction in $J mol^{-1}$:

$$\Delta_{surr}S = -\frac{\Delta H}{T}$$

Total entropy change and feasible reactions

If $\Delta_{tot}S$ is positive, at a certain temperature, then you can deduce that the reaction is feasible at that temperature.

Notice that if a feasible reaction is reversible, then the backwards reaction will have a negative $\Delta_{tot}S$ and will therefore not be feasible at this temperature.

If $\Delta_{tot}S = 0$ at a certain temperature then you can conclude that a reversible process will reach equilibrium where neither the forward or backward reaction is favoured.

Predicting the effect of changing temperature

Because changing temperature affects the magnitude of $\Delta_{tot}S$, it can also affect whether or not a reaction is feasible.

Worked example: Calculating a value for $\Delta_{tot}S$

Calculate $\Delta_{tot}S$ at 333 K for the reaction:

$$2NH_3(g) + 2O_2(g) \rightarrow N_2O(g) + 3H_2O(g) \qquad \Delta H = -552 \text{ kJ mol}^{-1}$$

$$H_2O(g) \qquad S^{\ominus} = +189 \text{ J K}^{-1} \text{mol}^{-1}$$

$$N_2O(g) \qquad S^{\ominus} = +220 \text{ J K}^{-1} \text{mol}^{-1}$$

$$NH_3(g) \qquad S^{\ominus} = +192 \text{ J K}^{-1} \text{mol}^{-1}$$

$$O_2(g) \qquad S^{\ominus} = +205 \text{ J K}^{-1} \text{mol}^{-1}$$

Step 1: Calculate $\Delta_{sys}S$ by adding up the molar entropies of reactants and products, remembering to multiply by the number of moles present in the equation:

$$\Delta_{sys}S = [220 + (3 \times 189)] - [(2 \times 192) + (2 \times 205)] = 787 - 794$$
$$= -7 \text{ J K}^{-1} \text{mol}^{-1}$$

Step 2: Calculate $\Delta_{surr}S$ using the equation $\Delta_{surr}S = \dfrac{-\Delta H}{T}$

$\Delta H = -552\,000$ J and $T = 333$ K, so $\Delta_{surr}S = \dfrac{-(552\,000)}{333} = \dfrac{552\,000}{333}$

$$= +1658 \text{ J K}^{-1} \text{mol}^{-1}$$

3 Add together the values of $\Delta_{sys}S$ and $\Delta_{surr}S$.

$$\Delta_{tot}S = (-7) + (1658) = +1651 \text{ J K}^{-1} \text{mol}^{-1}$$

Worked example: Predicting the sign of $\Delta_{sys}S$

Predict and explain the sign of $\Delta_{sys}S$ for this reaction:

$$Ca(s) + 2H_2O(l) \rightarrow Ca(OH)_2(s) + H_2(g)$$

Step 1: Compare the number of moles of gas on the two sides of the equation. In this equation there are 1 mole of gas on the RHS of the equation and none on the LHS.

Step 2: Decide whether entropy increases or decreases. There are more ways of arranging the particles in a gas than a solid, so the entropy increases.

Step 3: Relate this to the sign of $\Delta_{sys}S$: $\Delta_{sys}S = S$ (products) − S (reactants), so $\Delta_{tot}S$ is positive.

Revision tip

You will probably be given a value for ΔH in kJ mol^{-1}. Be sure to convert this into J mol^{-1}.

Revision tip

If you know that a reaction is feasible, you can work backwards to deduce that $\Delta_{tot}S$ is positive. This may in turn help you to deduce something about $\Delta_{sys}S$ or $\Delta_{surr}S$; for example, if you know that $\Delta_{surr}S$ is negative then $\Delta_{sys}S$ must be positive or the process could never be feasible.

Revision tip

You may be asked to describe what effect changing temperature has on the feasibility of a reaction, or to calculate the temperature at which a reaction becomes feasible.

Maths skill: Rearranging equations

To find T at which $\Delta_{tot}S$ is zero, you will need to be confident in rearranging equations.

Calculate the temperature in kelvin at which $\Delta_{tot}S$ is zero for the reaction:

$N_2(g) + 3H_2(g) \rightarrow 2NH_3(g)$ where $\Delta H = -92.4\,kJ\,mol^{-1}$ and $\Delta_{sys}S = -198.3\,J\,K^{-1}\,mol^{-1}$

$$0 = \Delta_{sys}S + \Delta_{surr}S$$

$$0 = -198.3 + -\frac{(-92\,400)}{T} = -198.3 + \frac{92\,400}{T}$$

1 Simplify the equation by moving $\Delta_{sys}S$ onto the LHS of the equation (add 198.3 to both sides):

$$198.3 = \frac{92\,400}{T}$$

2 Rearrange the equation so that T is no longer at the bottom of an expression (multiply both sides by T):

$$198.3T = 92\,400$$

3 Rearrange the equation to find T:

$$T = \frac{924\,00}{198.3}$$

$$T = 466.0\,K$$

Summary questions

1 Predict and explain the sign of $\Delta_{sys}S$ for this reaction:
   $N_2(g) + 3H_2(g) \rightarrow 2NH_3(g)$ (4 marks)

2 Calculate the value of $\Delta_{tot}S$ at 298 K for the decomposition of magnesium nitrate and comment on the significance of this value:
   $2Mg(NO_3)_2(s) \rightarrow 2MgO(s) + 4NO_2(g) + O_2(g)$ $\Delta H = +510.8\,kJ\,mol^{-1}$
   Values of S in $J\,K^{-1}\,mol^{-1}$: $Mg(NO_3)_2(s) = +164.0$, $MgO(s) = +26.9$, $NO_2(g) = +240.0$, $O_2(g) = +205.0$ (6 marks)

3 Magnesium carbonate is stable at room temperature, but decomposes when heated. Deduce the signs of $\Delta_{sys}S$, $\Delta_{surr}S$, and $\Delta_{tot}S$ at 298 K.   (4 marks)

Chapter 4 Practice questions

1 Which of these statements about entropy is true?

   A Entropy is a measure of the order of molecules and quanta.

   B Entropy is measured in units of $kJ\,mol^{-1}$.

   C The total entropy of the system and surroundings always increases if a reaction is feasible.

   D Exothermic reactions result in a decrease in total entropy. (*1 mark*)

2 Ammonia can be formed in the Haber process:

   $N_2(g) + 3H_2(g) \rightarrow 2NH_3(g)$ $\Delta H = -92.3\,kJ\,mol^{-1}$

   What can be concluded from this information about the entropy changes at 298 K for this process?

   1 The $\Delta_{sys}S$ is positive

   2 The $\Delta_{tot}S$ is positive

   3 The $\Delta_{surr}S$ is positive

   A 1, 2, and 3

   B 1 and 2

   C 2 and 3

   D only 3 (*1 mark*)

3 Which row of the table describes the difference in the properties of a sodium ion compared to a potassium ion?

	Ionic radius of Na⁺	Charge density of Na⁺	Λ_hyd H of Na⁺
A	Greater	Smaller	Greater
B	Smaller	Greater	Greater
C	Greater	Smaller	Smaller
D	Smaller	Greater	Greater

4 Which of these enthalpy changes are **always** exothermic?

   1 $\Lambda_{sol}H$

   2 $\Delta_{hyd}H$

   3 $\Delta_{LE}H$

   A 1, 2, and 3

   B 1 and 2

   C 2 and 3

   D only 3 (*1 mark*)

5 Sodium sulfate, Na_2SO_4, is soluble in water. A group of students measures the enthalpy change of solution of sodium sulfate. They dissolve a known mass of sodium sulfate in water and obtain a value of $-1.5\,kJ\,mol^{-1}$.

   a The standard enthalpy change of solution of sodium sulfate is $-2.5\,kJ\,mol^{-1}$.

   Suggest a reason why the value measured by the students is less negative than the databook value. (*1 mark*)

b **i** Complete this diagram to show the relationship between the enthalpy change of solution, lattice enthalpy and enthalpy change of solution for the ionic compound sodium sulphate.

Label the enthalpy changes and the details of the species involved.

(3 marks)

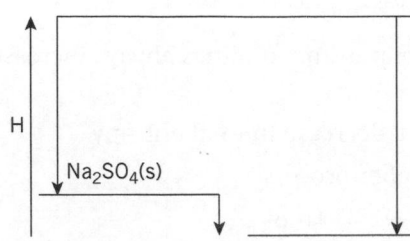

(5 marks)

ii The students find the following data in a databook:

$\Delta_{LE}H[Na_2SO_4] = -1944.0 \, kJ \, mol^{-1}$

$\Delta_{hyd}H[Na^+] = -406.0 \, kJ \, mol^{-1}$

Use these data, along with the standard enthalpy change of solution of sodium sulfate, to calculate a value for the standard enthalpy of hydration of the sulfate ion. *(4 marks)*

$\Delta_{hyd}H[SO_4^{2-}] = $ _____ $kJ \, mol^{-1}$

6 Many ionic compounds are soluble in water. However, calcium carbonate, $CaCO_3$, is usually regarded as an insoluble compound.

a Explain why many ionic compounds are soluble in water. *(3 marks)*

b An equation can be written to represent the dissolving of calcium carbonate:

$CaCO_3(s) + aq \rightarrow Ca^{2+}(aq) + CO_3^{2-}(aq)$ $\quad \Delta H = -13.0 \, kJ \, mol^{-1}$

i The entropy change for this reaction, $\Delta_{sys}S = -204.8 \, J \, K^{-1} \, mol^{-1}$

Calculate the value for the total entropy change, $\Delta_{tot}S$ at 298 K, and explain how this shows that calcium carbonate is insoluble at this temperature. *(3 marks)*

ii Discuss whether the dissolving of calcium could be made feasible by changing the temperature at which the dissolving process was attempted. *(3 marks)*

5.5 Protein structure

Proteins

Proteins are natural polymers, made from many amino acids bonded together by peptide links.

Levels of structure

Most proteins are large molecules with several levels of structure.

Primary structure

The primary structure is the **sequence of amino acids** in the protein chain. It is maintained by the covalent bonds in the peptide links that form when the amino group of one amino acid condenses with the carboxylic acid group in another amino acid.

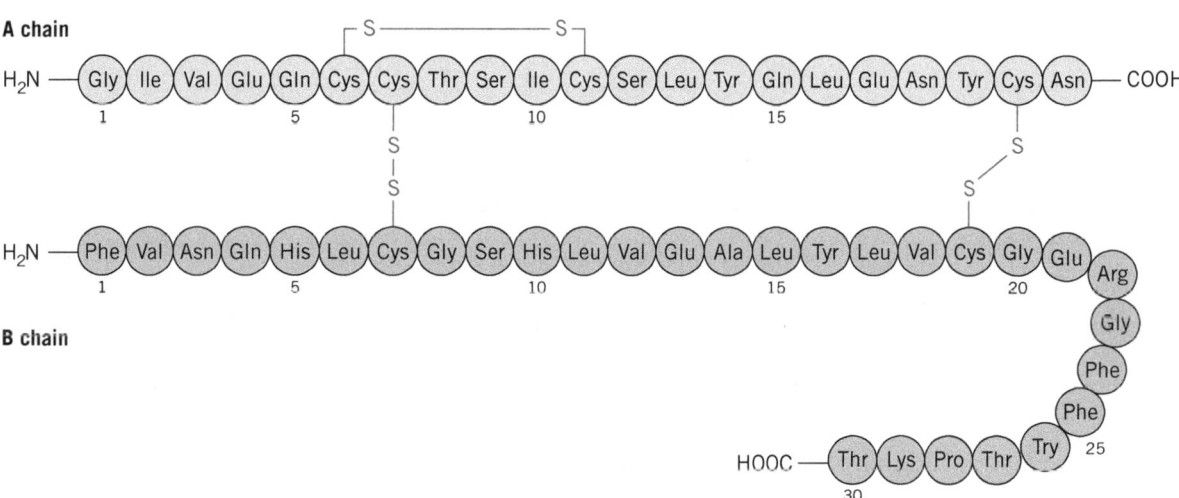

▲ **Figure 1** *The sequence of amino acids in the protein insulin*

Secondary structure

Secondary structure is the folding of the protein chain **into three-dimensional features (alpha helix, or beta pleated sheet).** The secondary structure is maintained by hydrogen bonds between the atoms in adjacent peptide links

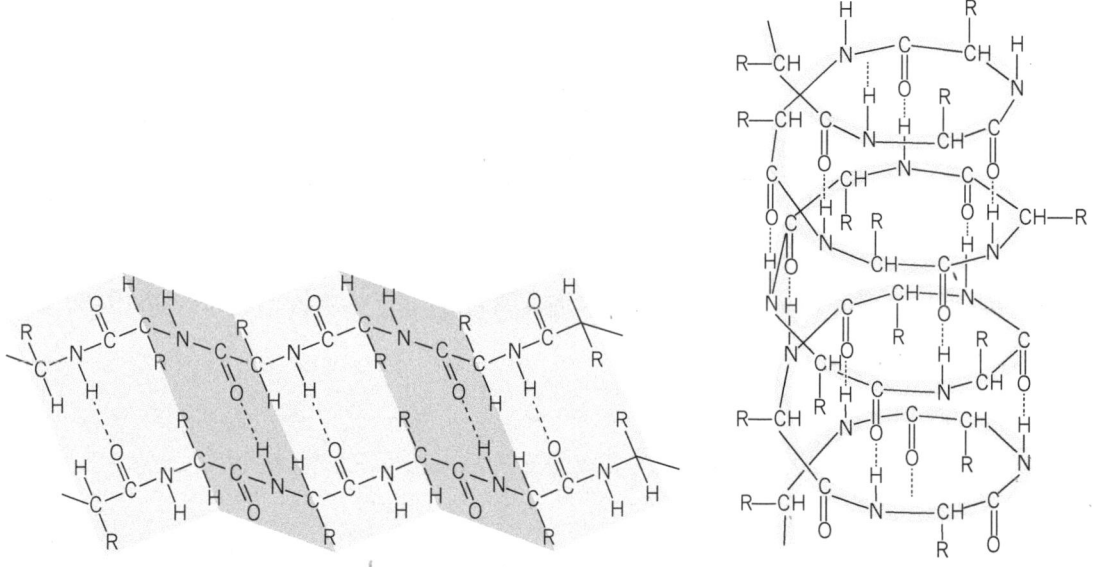

▲ **Figure 2** *Secondary structure within a protein is determined by hydrogen bonds between peptide links*

Go further

Regularly repeating primary sequence may allow polypeptide chains to pack together more closely; this means that the intermolecular bonds (strictly intramolecular bonds) are likely to be stronger; so more force is needed to allow the chains to slide past each other.

Tertiary structure

Further folding can occur, which causes the protein to take up a specific, complex three–dimensional shape. The folding happens because of the formation of bonds such as hydrogen bonds, ionic bonds, or instantaneous dipole–induced dipole bonds between the side groups of amino acids. One particular side group, on a cysteine amino acid, contains an –SH group, and these can form a covalent bond called a disulfide bridge.

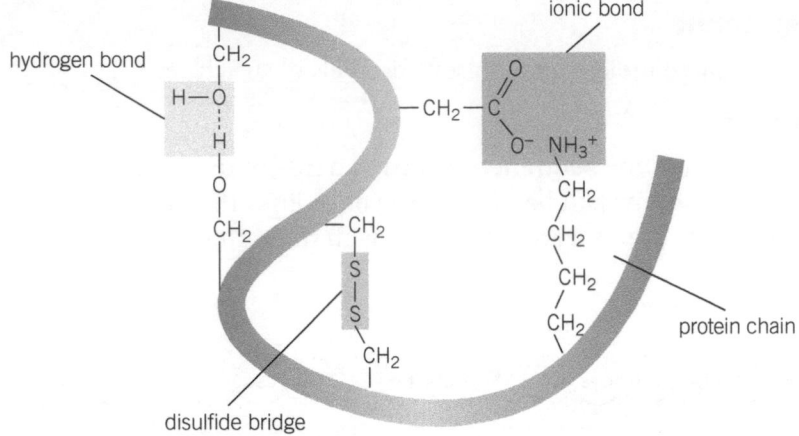

▲ **Figure 3** *The further folding to form a tertiary structure of a protein is determined by bonds between amino acid side groups*

Levels of structure and protein properties

The amino acid sequence (primary structure) determines the pattern of bonds that form with the protein and hence the secondary and tertiary structure. This in turn will determine the properties of a protein and the function of a protein in a biological system.

Enzymes always have a complex tertiary structure which includes a cleft called the **active site**. Small changes to the primary structure can make a big difference to the shape of the active site and its ability to bind to a substrate.

Proteins with a regularly repeating primary sequence will form extensive regions of regular secondary structure and may have a structural role in an organism such as in muscle fibres or hairs.

Summary questions

1 Describe the differences between the secondary and tertiary structure of a protein. *(4 marks)*

2 Draw a diagram to show how a hydrogen bond can form between two neighbouring peptide groups in a region of secondary structure. Indicate relevant lone pairs and partial charges. *(3 marks)*

3 Changes in the primary structure of an enzyme can result in it becoming inactive. Explain why. *(3 marks)*

5.6 Molecular recognition

Specification reference: PL (e)

Pharmacophores and receptor sites

Molecules that act as medicines possess structural features known as pharmacophores. These enable the molecule to bonds to a receptor site (or active site) in a target organism.

Some drug molecules act as **inhibitors** of enzymes.

Identifying pharmacophores

Chemists identify pharmacophores by examining the structures of several molecules that have similar biological actions. These molecules will have often have a structural feature in common: this will be the part of the molecule that binds to the receptor site and so it will be the pharmacophore.

Modifying the pharmacophore

Once chemists have identified the pharmacophore in a drug molecule, they will try and improve the properties of the drug (to make it more effective or have fewer side effects) by modifying the pharmacophore. They do this by changing the groups of atoms that are attached to the pharmacophore.

Interactions with receptor sites

Pharmacophores interact with receptor sites in a similar way to that in which a substrate binds to the active site of an enzyme.

The size and three-dimensional shape of a pharmacophore will be complementary to the shape of a particular receptor site. So pharmacophores can fit into, and bond to, that receptor site.

The functional groups on the pharmacophore will be oriented in a particular way that enables them to bond to groups in the receptor site by forming ionic bonds, hydrogen bonds, instantaneous dipole–induced dipole bonds, etc.

Many receptor sites contain chiral carbon atoms. If the pharmacophore is also chiral then only one enantiomer of the pharmacophore will have groups in the correct three-dimensional arrangement to bond to the receptor site.

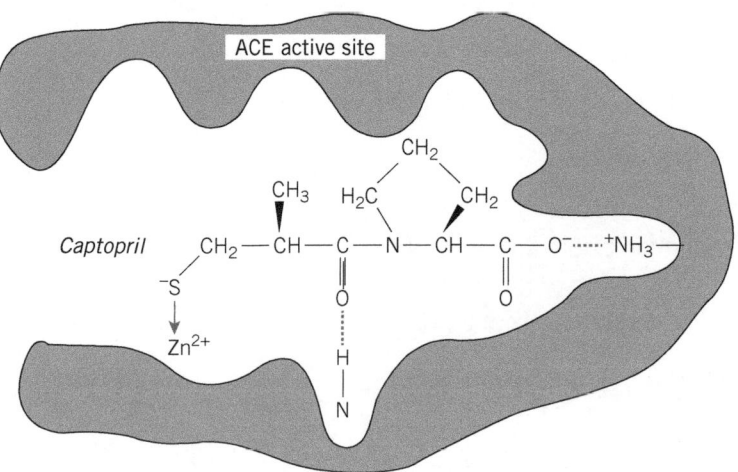

▲ **Figure 2** The pharmacophore in the drug captopril, binding to the active site of the enzyme ACE

Synoptic link

Active sites in enzymes are described in Topic 10.8, Enzymes.

Key terms

Pharmacophore: The part of a molecule that is responsible for its pharmacological activity.

Receptor site: These are found within certain proteins on the surface of cells. They recognise and bond to specific molecules in a similar way to the way that active sites bind to substrate molecules.

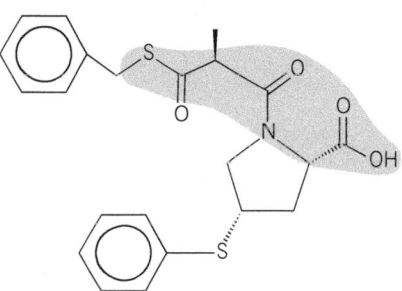

▲ **Figure 1** The drug zofenopril, showing the pharmacophore that is also found in other similar drugs

Key term

Inhibitor: A molecule that competes with a substrate molecule for the active site of an enzyme and prevents the substrate from binding.

Synoptic link

The action of enzyme inhibitors is described in Topic 10.8, Enzymes.

13

Go further

Examples of properties that might be altered by modifying a pharmocophore can include:

its solubility in water (blood) or fats (cell membranes)

its ability to bind to other, similar receptor sites

how quickly it is broken down in a living organism.

Question: Suggest how a pharmacophore could be modified to:

a make it more water soluble   **b** make it more soluble in fats.

Revision tip

You may be asked to identify the pharmacophore present in a drug molecule. Compare the molecule with the structure of other molecules that have the same biological effect, and find the largest part of the structure that all these molecules have in common.

🖩 Worked example: Enzyme inhibitors

The drug molecule captopril acts by inhibiting the enzyme ACE (see Figure 2 above). Explain how it does this.

- Captopril is likely to have a similar three-dimensional structure to that of the substrate of ACE.
- So it binds to the active site, competing with the substrate.
- This prevents substrate molecules bonding to the active site, so the enzyme becomes inactive.

Summary questions

1 **a** Explain what is meant by the term 'pharmacophore'.   (*1 mark*)
   b Describe one way in which chemists modify the structure of a pharmacophore.   (*2 marks*)

2 The two molecules below are both a type of penicillin. Suggest the structure of the pharmacophore in penicillin drugs.   (*1 mark*)

3 The drug thalidomide exists as two enantiomers. These enantiomers have different biological effects.
   a Explain why thalidomide exists as two enantiomers.   (*2 marks*)

   b Use ideas about receptor sites to explain why these enantiomers have very different biological effects.   (*2 marks*)

5.7 Bonding dyes to fibres

Specification reference: CD (a)

Colourfast dyes

Dyes' molecules possess a chromophore and are therefore coloured. However in order for them to be effective as dyes, they must also be able to attach strongly to fabrics or other solid substances. Dyes with this property are described as **colourfast**; it is difficult for the dye molecule to detach from the fibre when it is washed.

Dye–fibre interactions

Dyes can bond to fibres by making use of a range of interactions:

- intermolecular bonds, especially hydrogen bonding and instantaneous dipole–induced dipole
- ionic bonds
- covalent bonds, including dative covalent bonds to metal ions (mordants).

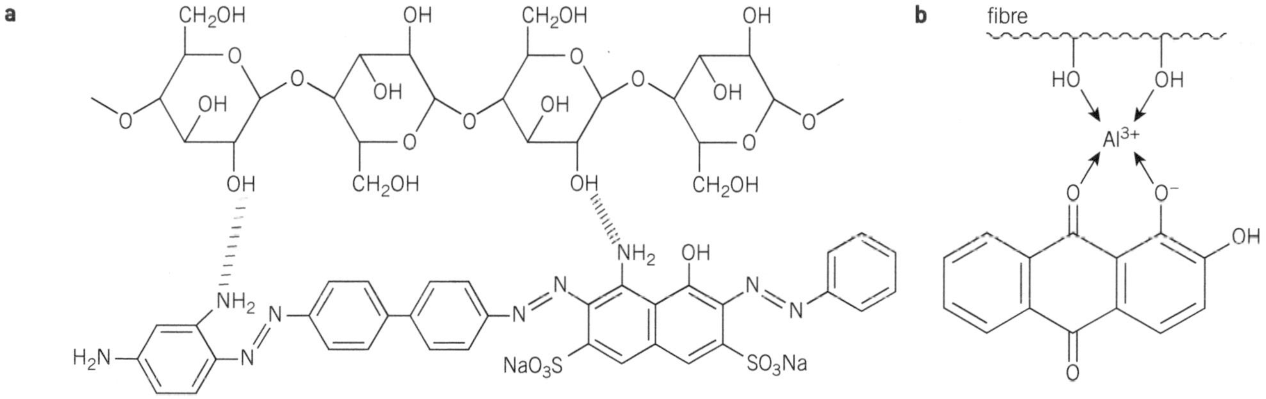

Revision tip

You can use ideas about the strength of bonds broken and formed to explain why dyes do not easily wash out of fibres. Think about the bonds present between the dye and the fibre and the bonds that would be possible between the dye and water molecules if the dye dissolved. This is similar to the way in which you would discuss why a solute may be insoluble in a particular solvent.

▲ **Figure 1 a** *OH groups on a dye molecule allow it to form hydrogen bonds to a cellulose fibre* **b** *an aluminium ion acts as a mordant by forming dative covalent bonds to both a dye molecule and a fibre*

Modifying the dye molecule

Certain groups are introduced into dye molecules in order to enable bonding to fibres:

- Acidic groups that ionise in aqueous solutions – the sulfonic acid group SO_3H is the most common acidic group.
- Fibre-reactive groups that can react in a condensation reaction with groups on the fibre.

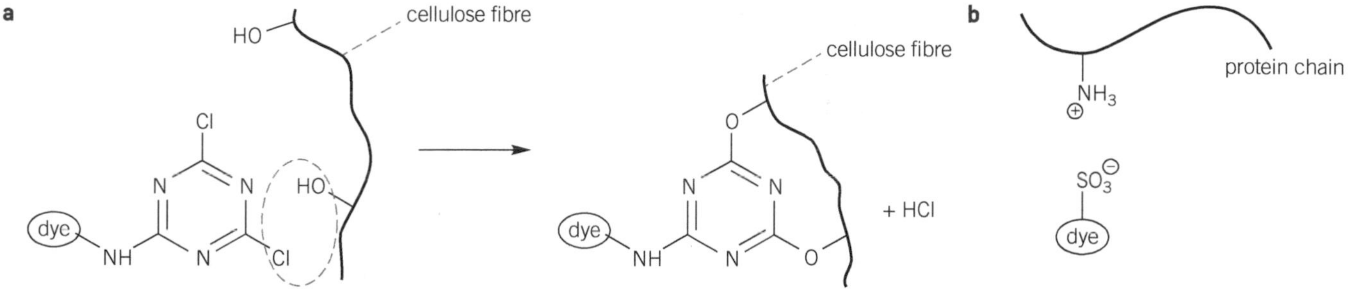

Synoptic link

The structure and synthesis of dyes is described in Topic 12.5, Reactions of arenes.

▲ **Figure 2 a** *a fibre-reactive dye forming a covalent bond to the amine group on a fibre* **b** *an acidic dye forming ionic bonds to a positively charged fibre*

Explaining the properties of dyes

A colourfast dye stays attached to the dye even when washed in warm water. You can use ideas about intermolecular bonding to explain why dyes are colourfast.

Model answer: Colourfast dyes

Disperse red is a non-polar dye and is used to dye polyester fibres. Explain why disperse red is a colourfast dye when in contact with water.

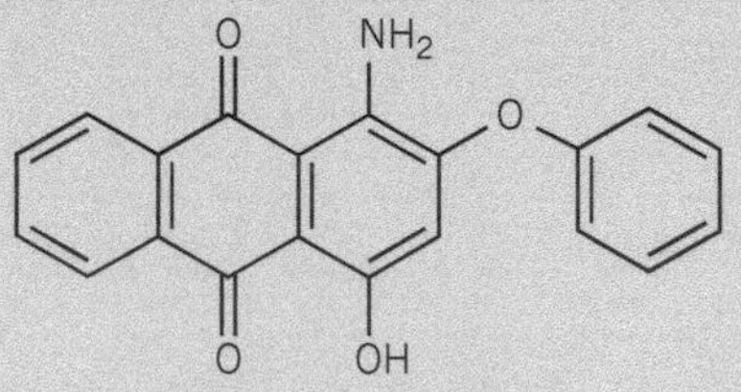

▲ **Figure 3** *Disperse red*

State the bonds present between the dye molecule and the fibres.

State the bonds present in the water.

State the bonds that can form between water and the dye molecule.

Compare the strength of the bonds described above.

Comment on the energy changes, and relate to the colourfast property of the dye.

- There are (strong) instantaneous dipole–induced dipole bonds between the dye molecule and the fibre.

- There are hydrogen bonds between water molecules.

- Some (weak) instantaneous dipole–induced dipole and a few hydrogen bonds can form between the dye molecule and water.

- The bonds that would need to break are stronger than the bonds that could form.

- The energy released from bond formation is not sufficient to compensate for the energy required to break bonds. So the dye will remain bonded to the fabric.

Synoptic link

Intermolecular bonds, such as hydrogen bonds and instantaneous dipole–induced dipole bonds, are described in Topic 5.3, Bonds between molecules: temporary and permanent dipoles.

Summary questions

1 List four types of bond that are used to bond dyes to fibres. (*4 marks*)

2 The dye acid blue contains a sulfonic acid group, SO_3H, and is used to dye nylon. The ends of nylon chains often contain basic NH_2 groups. Suggest how acid blue bonds to nylon. (*3 marks*)

3 Cotton (cellulose) can be dyed by using the colourfast dye direct red. Use ideas about the breaking and forming of bonds to explain why direct red is colourfast in water when it is used to dye cotton. (*6 marks*)

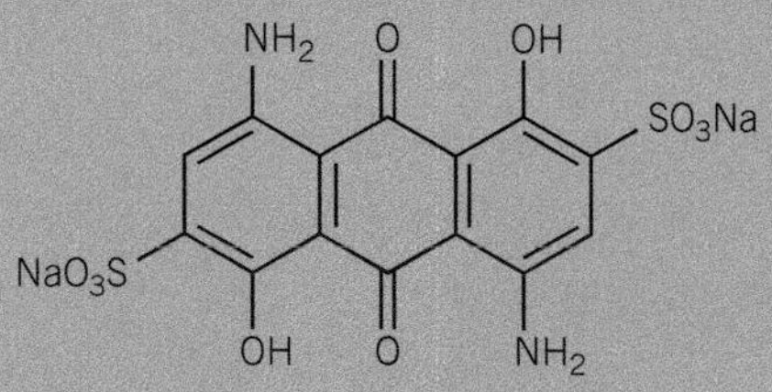

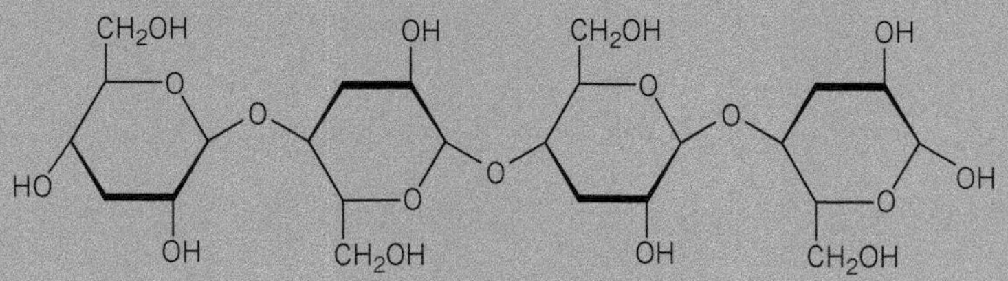

▲ **Figure 4** *The structures of* **a** *direct red and* **b** *cellulose*

1 Which of these statements about the secondary structure of proteins is correct?

   A It is maintained by covalent bonds within peptide groups.

   B It is maintained by hydrogen bonds between amino acid side chains.

   C It consists of regions of the protein that have specific three-dimensional shapes.

   D It has an important role in controlling the specificity of enzymes. (*1 mark*)

2 The tertiary structure of a protein is maintained by several types of bond. Which type of bond is most likely to be disrupted when the pH of a protein is altered.

   A Covalent bonds   C Instantaneous dipole–induced dipole

   B Ionic bonds   D Hydrogen bonds   (*1 mark*)

3 Related drug molecules often contain a specific pharmacophore. Which of these statements about pharmacophores is correct?

   1 The pharmacophore is the smallest section of a structure that is common to all the drug molecules.

   2 All pharmacophores are chiral molecules.

   3 If the structure of a pharmacophore is modified it may become even more effective.

   A 1,2, and 3   B 1 and 2   C 2 and 3   D only 3   (*1 mark*)

4 Dyes can bond to fibres by means of a range of intermolecular bonds.

   a Acid Blue dyes bond to wool fibres by ionic bonds.

      i Name the group of atoms in this structure responsible for the formation of ionic bonds to wool. (*1 mark*)

      ii Identify a group of atoms present in a protein structure that could form ionic bonds to this group. (*1 mark*)

   b Disperse dyes form instantaneous dipole–induced dipole bonds with polymer chains.

      i Describe how instantaneous dipole–induced dipole bonds form between two molecules. (*3 marks*)

      ii Suggest features in a dye molecule that will encourage the formation of relatively strong instantaneous dipole–induced dipole bonds. (*2 marks*)

   c Direct dyes often have hydroxyl groups which can hydrogen bond to alcohol or amine groups in fibres. Draw a diagram to show how a hydroxyl group can hydrogen bond to a primary amine group. Indicate partial charges and relevant lone pairs. (*4 marks*)

5 Dextroamphetamine is a drug used to treat disorders of the nervous system. The structure of dextroamphetamine is shown. It is thought that dextroamphetamine forms hydrogen bonds to receptor sites on the surface of nerve cells.

   a i Name the functional group in the dextroamphetamine molecule involved in binding to the receptor site. (*1 mark*)

      ii Explain how this group of atoms is able to form hydrogen bonds to the receptor site. (*2 marks*)

   b Dextroamphetamine exists as a pair of stereoisomers.

      i Explain why it can exist as stereoisomers. (*2 marks*)

      ii Draw suitable diagrams to show how the structures of these stereoisomers are related. (*2 marks*)

6.6 Further mass spectrometry

Specification reference: PL(r)

Synoptic link

The terms m/z and M⁺ peak were explained in Topic 6.5, Mass spectrometry.

Deducing molecular formulae and structures

The m/z value of the M⁺ peak gives you the relative molecular mass of a molecule. However it is not possible to deduce the molecular formula without further information, which can be provided by the technique of high-resolution mass spectrometry.

The fragmentation pattern obtained by low-resolution mass spectrometry can then be used to deduce the likely structure of a molecule.

High resolution mass spectrometry

High resolution mass spectrometry measures the mass of the M⁺ peak to four decimal places.

With this level of precision, it turns out that the relative mass of individual atoms are not whole numbers ($^{14}N = 14.0031$, $^{1}H = 1.0078$, $^{16}O = 15.9949$). ^{12}C however, is defined as exactly 12.0000.

This means that each molecular formula has a unique relative mass, to 4 d.p., e.g.

$$C_2H_6O = 46.0417$$

The mass of the M⁺ ion can be compared to a database and the molecular formula can be deduced.

Using the fragmentation pattern

A combination of the mass of the M⁺ ion and other information can tell you the molecular formula of the molecule.

However, there may be several isomers with the same molecular formula. The fragmentation pattern helps you to deduce which isomer is actually present.

From a knowledge of the molecular formula of the molecule and the mass of the fragment ions, you can deduce:

- The molecular formula of the fragment ions.
- The molecular formula of the other fragment that has been lost in order to form the fragment.

It may then be possible to deduce the structure of the molecule.

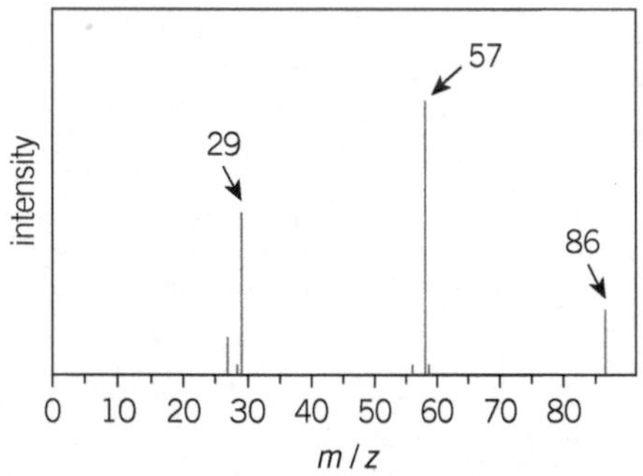

▲ **Figure 1** *The low resolution mass spectrum of the ketone $C_5H_{10}O$*

Revision tip

Remember that if a fragment is shown in the mass spectrum then you must write its formula with a positive charge. The groups of atoms that are lost from these fragment ions are uncharged and so do not show up in the spectrum.

 Worked example: Interpreting a mass spectrum

What information can be deduced from the mass spectrum of the ketone with molecular formula $C_5H_{10}O$, shown in Figure 1 above?

Step 1: Identify the molecular ion peak and use it to find the M_r value of the molecule: the M⁺ peak has the largest m/z value, i.e. 86. The M_r value is therefore 86, which is consistent with the molecular formula given.

Step 2: Identify the fragments that give rise to other peaks in the spectrum, using the difference between the M⁺ peak and the fragment peak, or by comparing the fragment peak mass with a list of common masses.

Peak at 57: 86 − 57 = 29, therefore a C_2H_5 *group* has been **lost**. The other alternative is CHO, but this group of atoms is only found in aldehydes, not ketones. The *fragment* is $C_3H_5O^+$.

Peak at 29: 29 suggests the presence of a $C_2H_5^+$ *fragment*. This suggests a *group* with m/z = 57 has been **lost**, which is likely to be C_3H_5O.

Small peak at 28: There is also a small peak at 28, which could be due to a CO^+ *fragment*. 86 − 28 = 58, which is 2 × 29, so maybe this is formed by the **loss** of two C_2H_5 *groups*.

Step 3: Use the information to suggest a structure for the molecule: the molecule is a ketone and also contains at least one C_2H_5 group. There do not seem to be any fragments suggesting larger alkyl groups (e.g C_3H_7) and so the most likely structure is $CH_3CH_2COCH_2CH_3$, pentan-3-one.

▼ **Table 1** *Common masses of species*

mass	possible formula
15	CH_3
17	OH
18	CO
29	C_2H_5 or CHO
43	C_3H_7 or CH_3CO
45	C_2H_5O or COOH
77	C_6H_5

Common misconception: Charges on fragments

Remember that the fragments that are detected as peaks in the mass spectrum will have a positive charge, even if this is not the usual charge for this group of atoms. So a fragment with a *m/z* value of 17 will be OH^+ not OH^-.

Summary questions

1   Explain how high-resolution mass spectrometry enables chemists to find the molecular formula of a molecule.   *(4 marks)*

2   The molecule butanone, C_4H_8O, has a structure $CH_3CH_2COCH_3$. The mass spectrum of butanone has fragment peaks at *m/z* = 57, 43, and 29.
    a   Explain how each of these fragments has formed from the molecular ion.   *(3 marks)*
    b   Give the formula of the species that causes each of these peaks.   *(3 marks)*

3   The molecules propyl methanoate ($HCOOC_3H_7$) and ethyl ethanoate ($CH_3COOC_2H_5$) are isomers. Discuss the value of high- and low-resolution mass spectrometry to enable a chemist to distinguish between these molecules.   *(6 marks)*

6.7 Nuclear magnetic resonance (NMR) spectroscopy

Specification reference: PL (s)

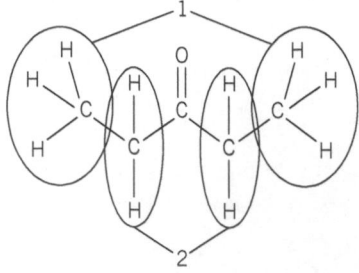

▲ **Figure 1** *Pentan-3-one has two different proton environments*

Key terms

Chemical shift (δ): A measure of the difference between the frequency absorbed by the protons in a peak and that absorbed by the protons in a reference compound (tetramethylsilane, TMS).

Multiplet: A peak that is split into several peaks, each with a very slightly different chemical shift.

Nuclear magnetic resonance spectroscopy is based on the fact that certain nuclei, such as ^1H and ^{13}C, have magnetic properties and can occupy high energy or low energy states when they are placed in a magnetic field.

The energy gap between these states corresponds to the absorption or emission of a photon of radio frequency electromagnetic radiation.

Proton (^1H) NMR spectroscopy

This provides information about the **chemical environment** of each type of proton in a molecule.

Chemical environment

The chemical environment of a proton is determined by factors such as how close it is in space to polar atoms, such as the O atoms in a C=O bond.

Proton (^1H) NMR spectra

Proton NMR spectra can either be low-resolution (low-res) or high-resolution (high-res).

Low-resolution NMR spectrum

In a low-resolution spectrum you will see a number of separate peaks at different chemical shifts.

From a low-resolution NMR spectrum you can deduce three pieces of information:

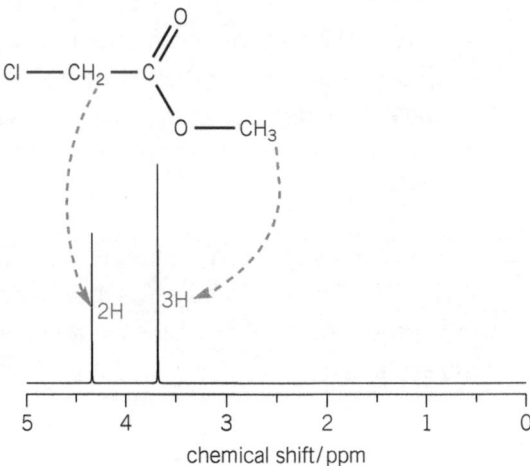

- The number of proton environments from the number of peaks.
- The relative number of protons in each environment from the height of the peaks (or strictly the area of the peaks) (although this is often indicated by a label on the peak such as '3H (or just 3)' etc.

▲ **Figure 2** *A low-resolution NMR spectrum of methyl chloroethanoate. There are two peaks showing that there are two different proton environments in this molecule. The ratio of protons in each environment is 2:3*

- Some information about the environment that produces each peak, from the value of the chemical shift.

^1H NMR chemical shifts relative to TMS

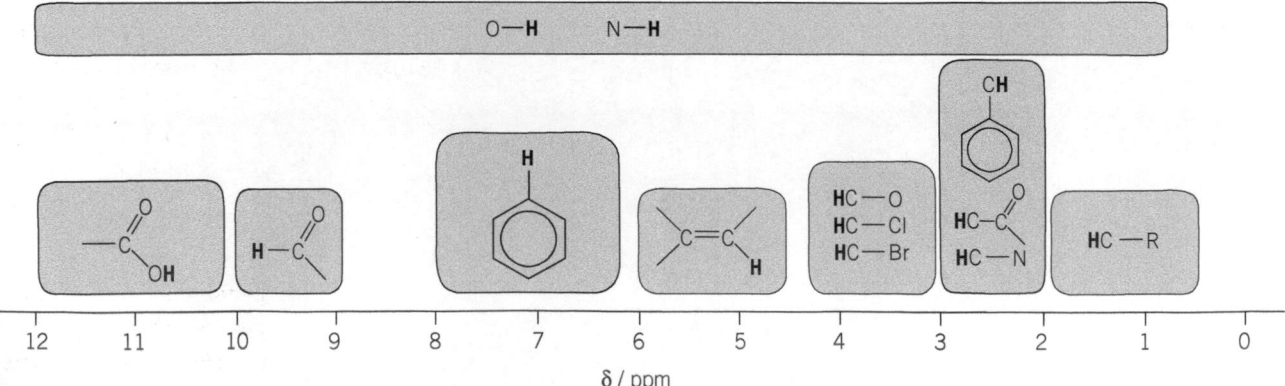

▲ **Figure 3** *The ^1H chemical shift data included in the data sheet*

High-resolution proton NMR spectra

In a high-resolution NMR spectrum, many of the peaks are shown to be split to form **multiplets**, such as doublets, triplets, or quartets. The pattern of this splitting provides more information about the groups of atoms that are adjacent to each proton environment.

Splitting patterns and the *n* + 1 rule

To predict the pattern of splitting observed for the peak caused by a proton environment, you use the *n* + 1 rule:

"The number of peaks in the multiplet caused by a proton environment equals n + 1, where n is the total number of equivalent protons on adjacent carbon atom(s)."

Go further: Coupling

Splitting arises because the magnetic field experienced by the protons in one environment can be affected by the alignment of the protons in an adjacent environment. Each proton can be aligned with the field ($\uparrow$) or against the field ($\downarrow$). So if there are two protons in a second, adjacent, environment these can be aligned in the following patterns:

$\uparrow\uparrow$ $\uparrow\downarrow$ $\downarrow\uparrow$ $\downarrow\downarrow$

Since $\uparrow\downarrow$ and $\downarrow\uparrow$ have the same effect, this means that there will be three different patterns of alignment, in the ratio 1:2:1. This causes the peak from the first environment to be split into a triplet, with the ratio of heights following the same 1:2:1 pattern.

Question: Use the same ideas of magnetic alignments to: **a** show that 3 protons in an adjacent environment will split a peak into a quartet; **b** predict the ratios of the peaks in the quartet.

Common misconception: OH and NH groups and splitting patterns

At this level, you only need to consider splitting caused by protons on adjacent C atoms.

You need to remember that the protons on OH, NH, or NH_2 groups do not cause splitting.

So even if there is a H attached to an O that is adjacent to a CH proton environment the peak due to this environment is not split.

Peaks due to OH or NH_2 groups are never split.

Carbon-13 (^{13}C) NMR spectroscopy

About 1% of the C atoms in a sample of an organic compound are in the form of a ^{13}C isotope. The nucleus of ^{13}C has magnetic properties, so a ^{13}C NMR spectrum can be obtained from every organic compound.

Revision tip

The peaks in the ^{13}C spectra that you will interpret will not be split into multiplets. Because only 1 in 100 atoms of C are ^{13}C isotopes it will be very unlikely that there will be two ^{13}C atoms adjacent to each other.

The presence of ^{1}H atoms does not cause splitting either because the spectra are obtained using a 'proton decoupling' technique.

Revision tip

You can use the data sheet to find the information about the likely chemical shifts of various common types of environment. Treat this information cautiously, as the actual chemical shift varies quite a lot depending on other factors. In general it is best to use chemical shift values to confirm the structure that you have deduced using other data, rather than starting with the chemical shift data.

Revision tip

You should make it clear how you have applied the *n*+1 rule in your answer. For example a triplet peak is caused by the presence of 2 protons on the carbon atom adjacent to the proton environment responsible for this peak.

You will not be expected to interpret multiplets more complex than a quartet (a peak split into 4).

Revision tip

You may be asked to predict the ^{1}H or ^{13}C spectrum of a compound. You should include information about the number of peaks and their likely chemical shifts. For ^{1}H spectra you should also predict the relative number of protons responsible for each peak and any splitting that might be seen for each peak.

Revision tip

Certain patterns of splitting are good evidence for particular groups of atoms. If there is a quartet and a triplet in the high-res ^{1}H NMR spectrum of a molecule, then this is good evidence for the presence of an ethyl group ($-CH_2-CH_3$).

The two protons in the CH_2 group split the peak for the adjacent CH_3 group into a triplet, and the three protons in the CH_3 group split the peak for the CH_2 group into a quartet.

Interpreting a ^{13}C spectrum

From a ^{13}C NMR spectrum you can deduce two pieces of information:

- The number of different environments of C atoms in the molecule – from the number of peaks.
- Some information about these environments, such as the group that the C is part of – from the chemical shift of each of these peaks.

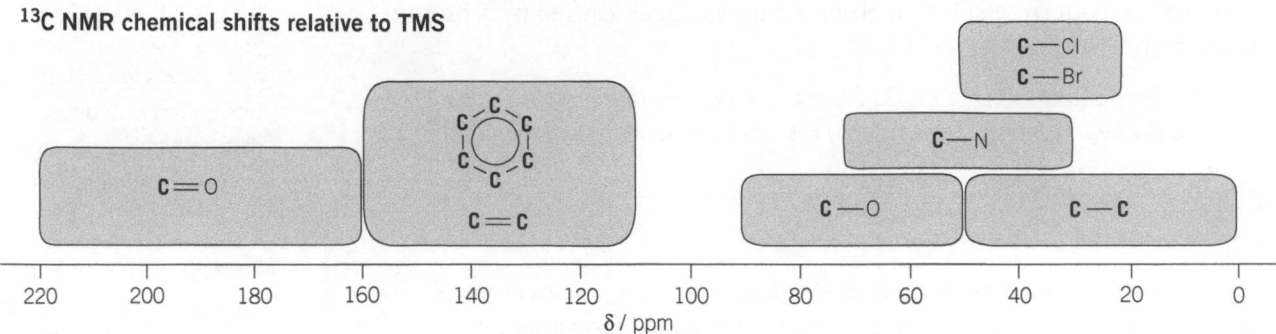

^{13}C NMR chemical shifts relative to TMS

▲ **Figure 4** *Chemical shifts of different 13C environments as shown in the OCR data sheet*

🖩 Worked example: Interpreting a ^{13}C NMR spectrum

The ^{13}C spectrum of a molecule with the molecular formula C_3H_6O is shown below:

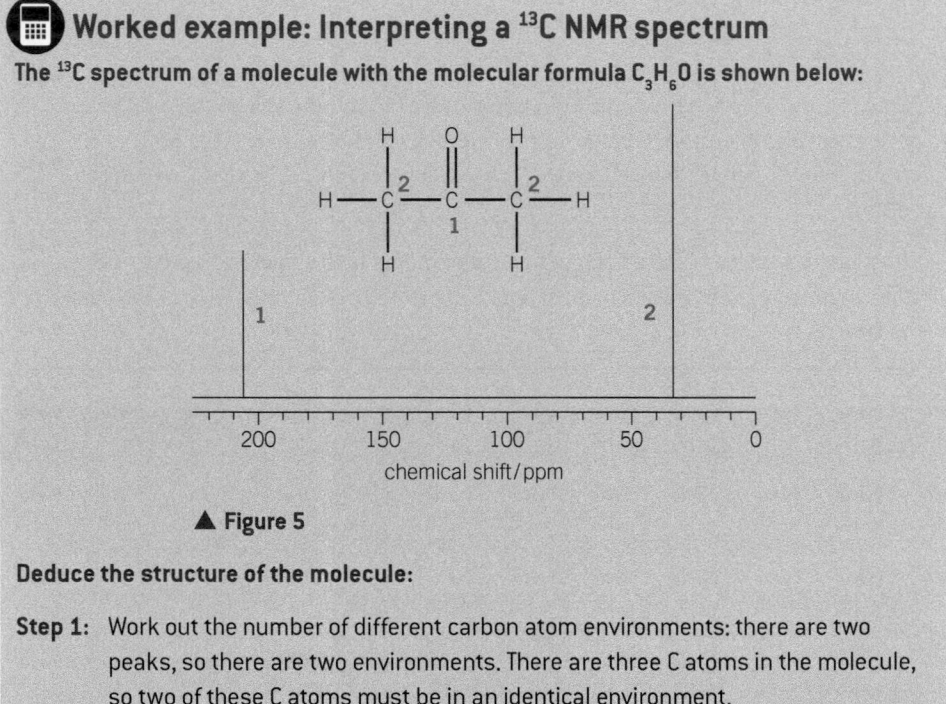

▲ **Figure 5**

Deduce the structure of the molecule:

Step 1: Work out the number of different carbon atom environments: there are two peaks, so there are two environments. There are three C atoms in the molecule, so two of these C atoms must be in an identical environment.

Step 2: Use the approximate ranges of chemical shifts to get an idea about the atoms that these C atoms are bonded to: $\delta = 207$ suggests C=O; $\delta = 32$ suggests C—C.

Step 3: Consider the possible structures of molecules with the formula C_3H_6O and explain which one matches the ^{13}C evidence most closely: Possible structures are propanone (CH_3COCH_3) or propanal (CH_3CH_2CHO). Propanal has three C environments, propanone has two C environments, so the molecule must be propanone. There are C=O and C—C groups in propanone, so this helps to confirm the choice.

Interpreting NMR spectra of aromatic compounds

The ^1H and ^{13}C nuclei in the molecule benzene are all in identical environments, so only one peak will be observed in the low-resolution ^1H or ^{13}C NMR spectrum of benzene.

If there are other groups substituted onto the benzene ring, then there may be several similar (but non-identical) environments.

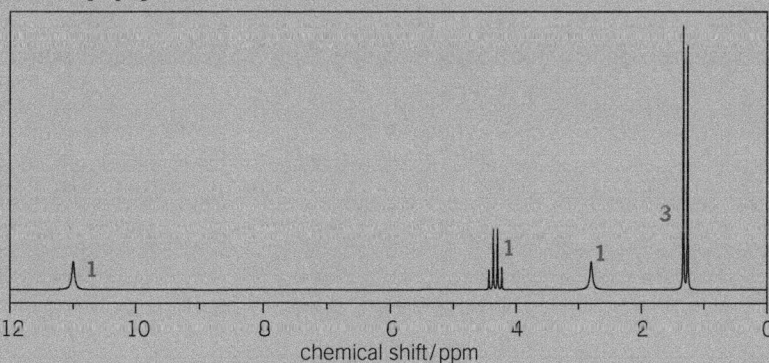

a 2H environments and
4C environments

b 3H and 4C environments

▲ **Figure 6** *The different environments of H and C atoms in two substituted benzene compounds*

Summary questions

1 Predict the number of peaks in **a** the low-resolution ^1H NMR spectrum and
b the ^{13}C NMR spectrum for the molecule shown below: *(2 marks)*

2 Explain why the molecule ethanal (CH_3CHO) produces a high-res ^1H
spectrum consisting of a quartet and a doublet group.   *(2 marks)*

3 The high-resolution ^1H NMR spectrum of a molecule with molecular
formula $C_3H_6O_3$ is shown below.

chemical shift/ppm

  a Deduce the structure of this molecule.   *(5 marks)*
  b Predict the appearance of the ^{13}C spectrum of the molecule. *(4 marks)*

6.8 Using combined spectroscopic techniques to deduce the structure of organic molecules

Specification reference: PL (t)

Information from different spectroscopic techniques

Different spectroscopic techniques provide different pieces of information that enable chemists to deduce the structure of an unknown compound. The empirical formula can be worked out from the % mass data. The M^+ peak from the mass spectrum gives the M_r so the molecular formula can be worked out from the empirical formula. (Alternatively the molecular formula can be worked out directly from the M^+ peak in the high-resolution mass spectrum.)

▼ **Table 1** *Information from different spectroscopic techniques*

Technique	Information	How to deduce the information
Mass spectrometry	the relative molecular mass	from the M^+ ion
Mass spectrometry	the groups of atoms present in the molecule	from the fragmentation pattern
High-resolution mass spectrometry	the molecular formula	from the mass of the M^+ peak (to 4 d.p.)
Low-res ^1H spectroscopy	the number of proton environments	from the number of peaks
Low-res ^1H spectroscopy	relative number of the protons in each environment	from the height of the peaks (or of the integrated trace)
Low-res ^1H spectroscopy	the types of proton environments	from the chemical shift of each peak
High-res ^1H spectroscopy	the number of H atoms on C atoms adjacent to each proton environment	from the splitting pattern of each peak
^{13}C spectroscopy	the number of ^{13}C environments	from the number of peaks
^{13}C spectroscopy	the types of ^{13}C environments	from the chemical shift of each peak
Infrared spectroscopy	the functional groups present in the molecule	from the wavenumbers of the peaks

The molecular formula can also be deduced from percentage by mass data. This enables you to work out the empirical formula. You can then use the relative molecular mass to find the molecular formula.

Synoptic link

Working out an empirical formula from percentage by mass data is explained in Topic 1.1, Amount of substance.

Synoptic link

The use of the M^+ peak to determine M_r values is described in Topic 6.5, Mass spectrometry.

Revision tip
You may be asked to explain how you used the data provided to deduce the structure of the molecule. Spend some time working out the structure in rough first. It is often easier to begin your answer by drawing out the structure first and then describe how the data links with the structure you have drawn.

Using combined spectroscopic techniques to identify a molecule

An unknown molecule has the empirical formula C_3H_4O. The M^+ peak in the mass spectrum occurs at an m/z value of 112. The infrared spectrum shows a peak at $1725\,cm^{-1}$, but no peak above $3100\,cm^{-1}$ or in the region $1600–1725\,cm^{-1}$. The ^1H NMR spectrum shows just one peak, at $\delta = 2.7$ ppm. In the high-res spectrum this peak is split into a triplet. The ^{13}C NMR spectrum has two peaks, at $\delta = 36$ and 210 ppm. You can use this information to deduce the structure of the molecule.

Step 1: Find the molecular formula of the molecule: The M_r of a C_3H_4O formula would be 56. The M_r of the actual molecule, from the M^+ peak, is 112. So there are two C_3H_4O units in the molecular formula, which is therefore $C_6H_8O_2$.

Step 2: Deduce the functional groups present in the molecule: There are no O–H bonds present (no peak above 3100) or C=C bonds (no peak between 1600–1700 cm^{-1}). However the peak at 1725 cm^{-1} indicates a C=O in either an aldehyde or a ketone; it can't be a carboxylic acid because there are no O–H bonds present. There must be two identical C=O groups, because there is only one peak.

Step 3: Deduce the number of ^1H and ^{13}C environments: There is just one peak in the ^1H NMR spectrum so all the ^1H nuclei must be in the same environment. There are two peaks in the ^{13}C spectrum so there are just two different ^{13}C environments. This strongly suggests that the molecule must be quite symmetrical.

Step 4: Use the splitting of the ^1H peaks to suggest some structural features present in the molecule: The presence of triplets suggest that there are CH_2 groups adjacent to each ^1H environment. Because the four ^1H environments are identical, there must be two $CH_2–CH_2$ groups present.

Step 5: Put the information together to suggest a structure: two ketone groups (C=O) and two $CH_2–CH_2$ groups. The only structure which works is Figure 1.

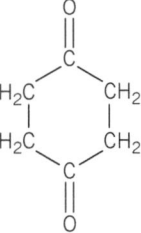
▲ Figure 1

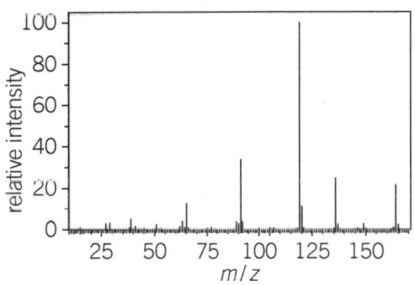
▲ Figure 2

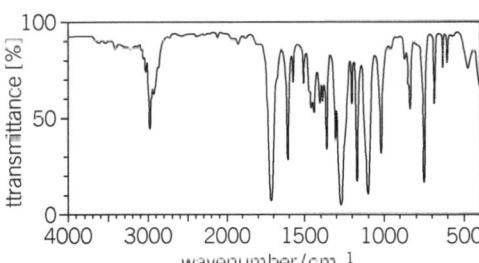
▲ Figure 3

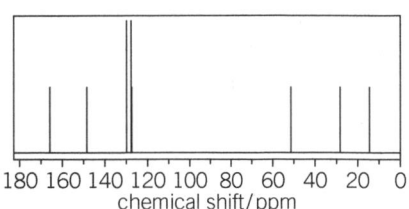
▲ Figure 4

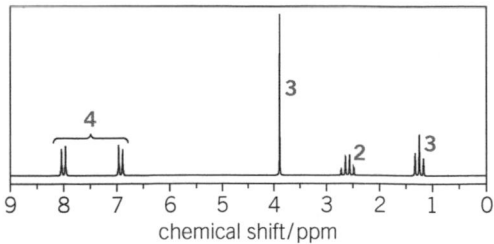
▲ Figure 5

Summary questions

1 Explain what information can be obtained from a combination of percentage by mass data and the M^+ ion peak from mass spectrometry. *(3 marks)*

2 Infrared spectroscopy and mass spectrometry can be used to distinguish between pairs of molecules. For each pair of molecules, state which method would allow these pairs of molecules to be distinguished most easily. For your chosen method, describe **one** way in which the results would be different for the two molecules.

   a

   i   ii

   b i $CH_2=CH–CH_2OH$ and **ii** CH_3CH_2CHO *(6 marks)*

3 An unknown compound has the following percentage composition by mass: C, 73.17%   H, 7.32%   O, 19.51%
   The mass spectrum, infrared spectrum, carbon-13, and proton NMR spectra are shown in Figures 2–5. Analyse this information to suggest the structural formula of the compound. *(6 marks)*

6.9 Coloured organic molecules

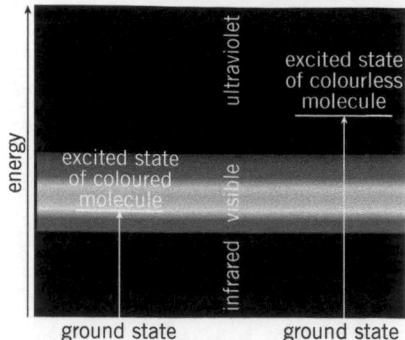

▲ **Figure 1** *The energy gap between energy levels in a molecule determines whether a molecule is coloured or colourless*

Key term

Complementary colour: Pairs of colours that are opposite to one another on a colour wheel, and which, when combined together, produce white light.

Electronic transitions and colour

When electromagnetic radiation from the visible or ultraviolet part of the spectrum is absorbed by molecules, electrons are excited from a lower energy level (the ground state) to a higher one (an excited state).

The frequency of radiation absorbed is related to the energy gap between these energy levels by the equation $\Delta E = h\nu$.

If the energy gap corresponds to a frequency of radiation in the range of visible light, the molecule will absorb some of the colours of visible light. The **complementary colour** will be transmitted or reflected and the molecule will appear coloured.

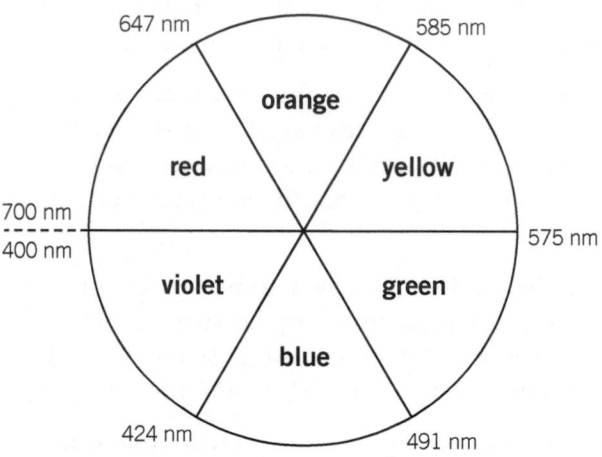

▲ **Figure 2** *The colour wheel*

Chromophores

The part of the molecule responsible for the absorption of visible light (or uv radiation) is known as the **chromophore**.

Many chromophores consist of a conjugated system in which there is extensive delocalisation of electrons. This occurs when there are alternating double and single bonds in a molecule.

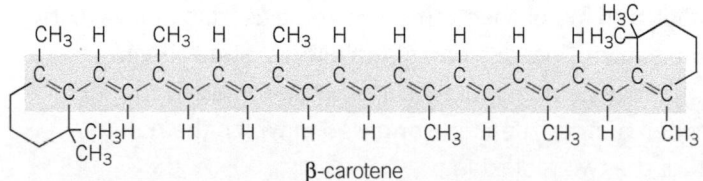

β-carotene

▲ **Figure 3** *The chromophore (shaded) in β-carotene consists of a conjugated system of alternating double and single bonds*

Synoptic links

The delocalisation of electrons in a benzene ring is described in Topic 12.3, Structural isomerism and E/Z isomerism.

The structure of dye molecules such as azo dyes is described in Topic 12.5, Reactions of arenes.

The relationship between chromophore and colour

The more extensive the delocalised system in a chromophore, the smaller the energy gap between energy levels.

Because a smaller energy gap corresponds to a lower frequency of light, molecules with extensive delocalised systems, such as azo dyes, absorb in the visible part of the spectrum.

Molecules with less extensive delocalisation, such as benzene, have a larger gap between energy levels and absorb a higher frequency of radiation. So benzene absorbs in the ultraviolet part of the spectrum.

Modifying the chromophore

Chemists modify the chromophore in a coloured molecule by adding groups that extend the delocalised system. This reduces the gap between energy levels and results in the molecule absorbing a lower frequency of radiation.

Synoptic link

Topic 12.5, Reactions of arenes, explains how the chromophore is modified in azo dye molecules.

Key terms

Chromophore: The part of an organic molecule responsible for absorbing visible light or ultraviolet radiation.

Conjugated system: A region of the molecule in which there is extensive delocalisation of electrons.

Summary questions

1 Dye molecules are organic molecules that possess a chromophore:
   a What is meant by the term 'chromophore'?
   b What structural feature is usually present in the chromophores of organic molecules? *(2 marks)*

2 The structure of the dye molecule disperse red 60 is shown below:

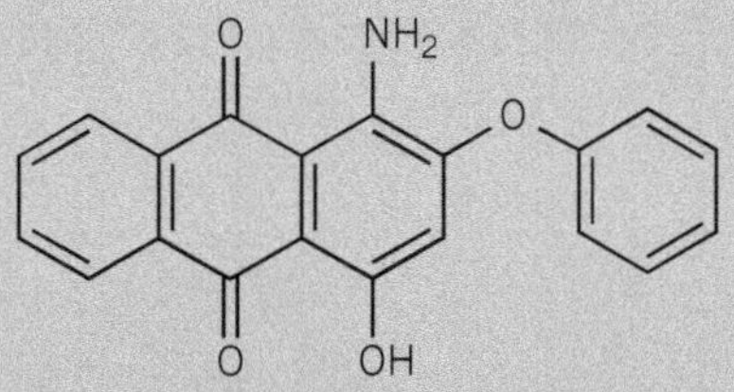

   Copy the structure and circle the chromophore in this molecule.

3 The structures of two molecules with different chromophores are shown below:

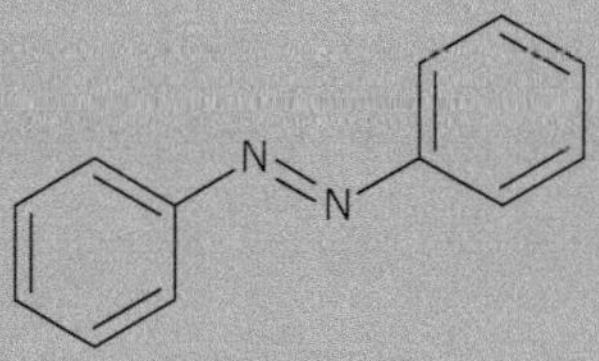

   a is orange, while b is colourless. Explain the difference in appearance of these two molecules. *(5 marks)*

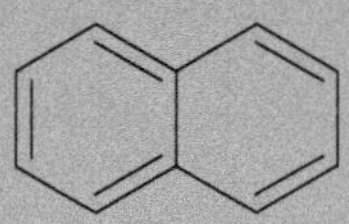

1 Which of these statements is true about high-resolution mass spectrometry?

   A The m/z value of the molecular ion enables the structure of the molecule to be determined.

   B The technique uses the fact that relative masses of isotopes are not exact integers.

   C Different isomers will have different masses when measured to 4 decimal places.

   D The technique can detect fragment species that do not have a charge.

   (1 mark)

2 A molecule has the molecular formula C_4H_8O. A peak is observed in the mass spectrum with a m/z value of 29.

   The species responsible for this peak could be:

   A CH_3CH_2

   B $C_4H_5^+$

   C CHO^+

   D C_2H_5O

   (1 mark)

3 A molecule has the structure shown below. Which row shows the correct number of environments in the ^{13}C and proton NMR spectra?

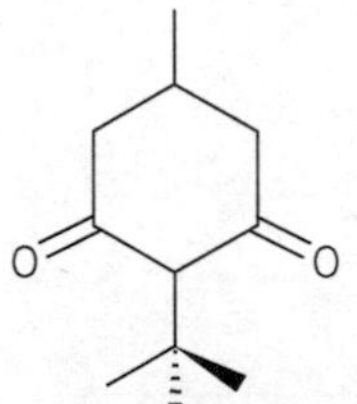

	Number of environments in the ^{13}C NMR spectrum	Number of environments in the proton NMR spectrum
A	7	7
B	10	10
C	9	6
D	7	5

4 In the proton NMR spectrum of propanal, CH_3CH_2CHO, the peak caused by the proton in the CHO environment will:

   1 Have a chemical shift of $\delta = 3.5\,ppm$

   2 Not be split

   3 Have an area less than any of the other peaks

   A 1, 2, and 3

   B 1 and 2

   C 2 and 3

   D Only 3

   (1 mark)

5 A molecule has the molecular formula $C_4H_8O_2$.

   The proton NMR spectrum of this molecule shows the following peaks.

Chemical shift / ppm	Relative number of H atoms	Splitting
1.2	3	triplet
2.0	3	singlet
4.2	2	quartet

   Suggest a structure for this molecule, explaining how the data from the spectrum links to the structure.

   (4 marks)

6 The structure of benzene and the dye molecule Disperse Red are shown below:

Benzene 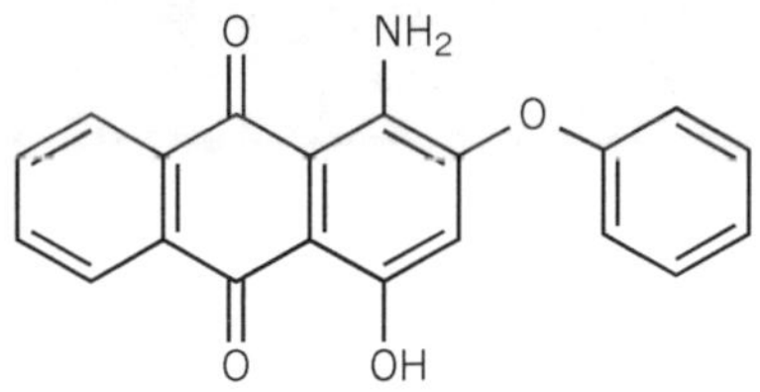   Disperse Red

Benzene is a colourless liquid whilst Disperse Red is a solid with an intense red colour when in suspension.

Use ideas about the bonding in these substances to explain the differences in colour of these two substances.   (*5 marks*)

7.5 Equilibrium constant K_c, temperature, and pressure

Specification reference: CI(f), CI(h)

Key term

Catalysts: Catalysts speed up a reaction without getting used up. They do this by providing a route of lower activation enthalpy.

Common misconception: The effect of catalysts

Remember that catalysts do not affect the position of equilibrium.

Equilibrium constant K_c

You have seen that K_c is a mathematical expression incorporating the equilibrium concentrations of each substance present, and that K_c is constant for a given reaction at a specified temperature:

$$K_c = \frac{[C]^c[D]^d}{[A]^a[B]^b}$$

Changing the pressure or concentration may affect the position of equilibrium and the equilibrium concentrations of the substances present, but the value of K_c remains constant.

Catalysts

Catalysts increase the *rate* of the reaction, so equilibrium is achieved more quickly, but adding a catalyst has no effect on the position of equilibrium or the numerical value of K_c.

Temperature

The *only* thing that affects the magnitude of K_c is a change in *temperature*:

- Increasing the temperature moves the position of equilibrium in the endothermic direction.
- Decreasing the temperature moves the position of equilibrium in the exothermic direction.

For example, in the Haber process, the following reaction takes place:

$$N_2(g) + 3H_2(g) \rightleftharpoons 2NH_3(g) \qquad \Delta H = -92\,kJ\,mol^{-1}$$

The forward reaction is exothermic, so increasing the temperature moves the position of equilibrium to the left. Therefore K_c decreases.

Decreasing the temperature moves the position of equilibrium to the right, in the exothermic direction, increasing the concentration of the product and increasing K_c.

▼ **Table 1** *The effects of temperature change*

	Exothermic reactions	Endothermic reactions
Temperature increases	value of K_c decreases	value of K_c increases
Temperature decreases	value of K_c increases	value of K_c decreases

Revision tip

Temperature is the only factor to alter K_c.

How does K_c change?

▼ **Table 2** *Changes that affect K_c*

Change	Position of equilibrium	K_c	Rate
Concentrations	changed	unchanged	changed
Total pressure	may change	unchanged	may change
Temperature	changed	changed	changed
Catalyst used	unchanged	unchanged	changed

Synoptic link

You will come across other equilibrium constants, such as K_a in connection with weak acids in Topic 8.2, Strong and weak acids and pH.

Units of K_c

Units of K_c can be calculated by substituting the units of concentration ($mol\,dm^{-3}$) for each species into the K_c expression, raising to a power ($mol^2\,dm^{-6}$, $mol^3\,dm^{-9}$ etc.) where appropriate. It is then necessary to cancel from the top and bottom of the expression.

 Worked example: The units of K_c (1)

$$K_c = \frac{[NH_3]^2}{[N_2][H_2]^3}$$

Calculate K_c when $[NH_3] = 2.25 \times 10^{-5}\,mol\,dm^{-3}$, $[N_2] = 2.49 \times 10^{-3}\,mol\,dm^{-3}$, and $[H_2] = 9.91 \times 10^{-3}\,mol\,dm^{-3}$.

Step 1: $K_c = \dfrac{(2.25 \times 10^{-5}\,mol\,dm^{-3})^2}{(2.49 \times 10^{-3}\,mol\,dm^{-3})\,(9.91 \times 10^{-3}\,mol\,dm^{-3})^3}$

Step 2: $K_c = \dfrac{5.0625 \times 10^{-10}\,mol^2\,dm^{-6}}{2.49 \times 10^{-3}\,mol\,dm^{-3} \times 9.73 \times 10^{-7}\,mol^3\,dm^{-9}}$

Step 3: $K_c = 0.21\,mol^{-2}\,dm^6$

In this case, the units on the top are $mol^2\,dm^{-6}$ and the units on the bottom are $mol^4\,dm^{-12}$. Cancelling from the top and bottom leaves $mol^{-2}\,dm^6$.

 Worked example: Calculating K_c from initial concentrations

$$K_c = \frac{[CH_3CO_2C_2H_5]\,[H_2O]}{[CH_3COOH]\,[C_2H_5OH]}$$

The initial concentrations of CH_3COOH and C_2H_5OH were $0.500\,mol\,dm^{-3}$. No $CH_3CO_2C_2H_5$ and H_2O were present in the initial mixture. At equilibrium, the concentrations of $CH_3CO_2C_2H_5$ and H_2O were $0.333\,mol\,dm^{-3}$. Deduce the equilibrium concentrations of CH_3COOH and C_2H_5OH and calculate the value of K_c including units.

Step 1: As the concentrations of $CH_3CO_2C_2H_5$ and H_2O increased to $0.333\,mol\,dm^{-3}$, the concentrations of CH_3COOH and C_2H_5OH will have *decreased* by $0.333\,mol\,dm^{-3}$, due to the 1:1 mole ratio. Therefore the equilibrium concentrations are $0.500 - 0.333 = 0.167\,mol\,dm^{-3}$.

Step 2: Substitute the concentrations into the expression for K_c.

$$K_c = \frac{(0.333\,mol\,dm^{-3})(0.333\,mol\,dm^{-3})}{(0.167\,mol\,dm^{-3})(0.167\,mol\,dm^{-3})}$$

Step 3: $K_c = 3.98$ (no units)

In this case, K_c has no units because $mol\,dm^{-3}$ cancels from the top and the bottom.

Techniques to determine equilibrium constants

To determine an equilibrium constant it is necessary to determine the concentrations of the substances involved. This can be done, for example, by titration if one of the substances is acidic. Then the concentrations of the other substances can be deduced from their mole ratio. Spectroscopic methods and calorimetric methods can also be used.

Key term

Stoichiometry: Stoichiometry refers to the mole ratio of reactants and products in a chemical equation.

Summary questions

1 a Write K_c for the reaction $CO(g) + Cl_2(g) \rightleftharpoons COCl_2(g)$ ($\Delta H = -108\,kJ\,mol^{-1}$).
      (*1 mark*)
  b Describe the effect on K_c of: (i) increasing pressure, (ii) increasing temperature, (iii) adding a catalyst.
      (*3 marks*)

2 For the reaction $N_2O_4 \rightleftharpoons 2NO_2$ ($\Delta H = +54\,kJ\,mol^{-1}$), explain the effect on K_c of increasing the temperature. (*3 marks*)

3 Calculate K_c, including units, for the reaction $H_2 + I_2 \rightleftharpoons 2HI$, given that $[H_2] = 0.004\,mol\,dm^{-3}$, $[I_2] = 0.004\,mol\,dm^{-3}$ and $[HI] = 0.027\,mol\,dm^{-3}$.
      (*4 marks*)

7.6 Solubility equilibria

Specification reference: O(h)

Key term

Solubility product: Solubility product is an equilibrium constant for the dissolving of a sparingly soluble salt.

Solubility product

K_{sp} is an equilibrium constant known as the solubility product.

A sparingly soluble ionic solid is in equilibrium with its constituent ions. For example:

$$AB(s) \rightleftharpoons A^+ (aq) + B^- (aq)$$

Because the concentration of a solid cannot be measured, the equilibrium constant for this reaction is the product of the concentrations of the two ions:

$$K_{sp} = [A^+][B^-]$$

 Worked example: Solubility product calculations

Will a precipitate form from a solution at 298 K in which $[Ag^+]$ and $[Cl^-] = 5.0 \times 10^{-6}$ mol dm^{-3}?

$K_{sp}(AgCl) = 2.0 \times 10^{-10}$ mol^2 dm^{-6} at 298 K

$[Ag^+]$ and $[Cl^-] = $
5.0×10^{-6} mol dm$^{-3} \times 5.0 \times 10^{-6}$ mol dm$^{-3} = 2.5 \times 10^{-12}$ mol^2 dm^{-6}

This answer is smaller than K_{sp}, so no precipitate will form.

 Worked example: Writing K_{sp}

Step 1: AgCl (s) $\rightleftharpoons$ Ag$^+$ (aq) + Cl$^-$ (aq)

$K_{sp} = [Ag^+][Cl^-]$

Step 2: Lead(II) iodide contains the ions Pb^{2+} and I$^-$ in a 2 : 1 ratio, so the expression for K_{sp} is:

$K_{sp} = [Pb^{2+}][I^-]^2$

Solubility product calculations

Given the value of K_{sp}, it is possible to predict whether a precipitate will form. If the product of the concentration of the ions (raised to a power as necessary) is *smaller than or equal to K_{sp}*, the ions will stay in solution and a precipitate will not form.

Techniques for determining solubility products

To determine a solubility product experimentally, it is necessary to measure the concentrations of the ions in solution. With Ca(OH)$_2$, for example, this can be done by titration to determine [OH$^-$]. It is necessary to filter the undissolved solid calcium hydroxide before titrating. The concentration of [Ca^{2+}] can then be deduced by the mole ratio of the reaction since [OH$^-$] will be twice as great as [Ca^{2+}].

Revision tip

A precipitate will not form if the product of the concentration of the ions is smaller than or equal to K_{sp}.

Summary questions

1 Write an expression for K_{sp} for:
  a AgBr
  b Ag$_2$S
  c Ag$_2$CrO$_4$ *(3 marks)*

2 Calculate K_{sp} for AgBr given that $[Ag^+]$ and $[Br^-] = 7.07 \times 10^{-7}$ mol dm^{-3}.
                                                               (3 marks)

3 Will a precipitate form from a solution at 298 K in which $[Pb^{2+}]$ and $[SO_4^{2-}] = 1.45 \times 10^{-4}$ mol dm^{-3}? $K_{sp}(PbSO_4) = 1.60 \times 10^{-8}$ mol^2 dm^{-6} at 298 K.   *(2 marks)*

7.7 Gas–liquid chromatography

Specification reference: CD(n)

Stationary phase and mobile phase

Chromatography is a method of separating and identifying the components of a mixture.

In chromatography, equilibrium is established as components are distributed between the **stationary phase** and the **mobile phase**. Components with a greater affinity for the stationary phase consequently move more slowly than those with a lower affinity.

Gas–liquid chromatography

In gas–liquid chromatography (GLC), the stationary phase is a non-volatile liquid coated on the surface of finely divided solid particles. This material is packed inside a long thin **column**, which is coiled inside an oven. A **carrier gas**, which is unreactive, is the mobile phase and carries the mixture through the column.

A peak is recorded on the **chromatogram**, as each component emerges from the column. The *time* that a component takes to emerge is called the **retention time** and the *area* under each peak is proportional to the amount of that component in the mixture.

When the components emerge from the column they may pass into a mass spectrometer, which gives further information about the structure of the component, including its relative molecular mass.

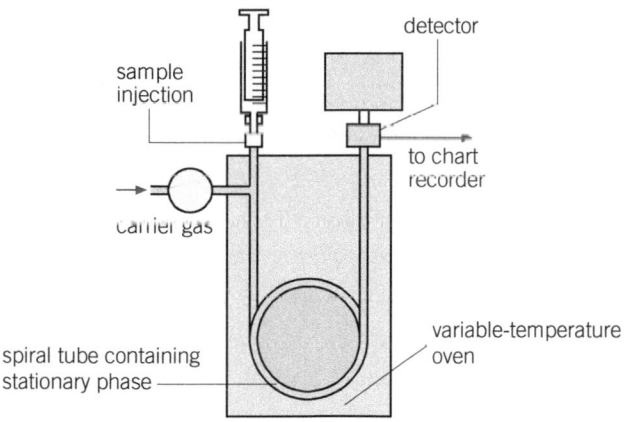

▲ **Figure 1** GLC apparatus

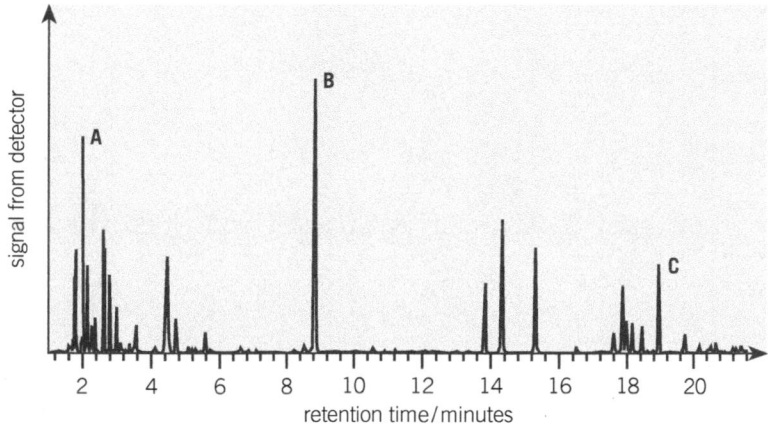

▲ **Figure 2** A typical gas–liquid chromatogram

Key term

Stationary phase: The material fixed in place in chromatography, for example a paper strip, the silica coating on a thin-layer chromatography plate, or the liquid on a solid support in gas–liquid chromatography.

Key term

Mobile phase: The substance which flows in a specific direction in chromatography, carrying the separating components with it.

Key terms

Chromatogram: The result of a chromatographic separation.

Retention time: The time taken for a component to emerge from the column in gas–liquid chromatography.

Revision tip

Nitrogen and noble gases are used as inert carrier gases.

Revision tip

An alternative to a packed column is a very long and very thin capillary tube with a high boiling point liquid coating on the inside surface.

Synoptic link

You learned about mass spectrometry in Topic 6.5, Mass spectrometry.

🔢 **Worked example: Interpreting a gas–liquid chromatogram**

Referring to the chromatogram in Figure 2, identify:

a   the component which makes up the greatest proportion of the mixture,

b   the retention time of peaks A, B, and C,

c   which component, A, B, or C has least affinity for the mobile phase.

a   Component B is the tallest peak so it makes up the greatest proportion of the mixture. Because the peaks are very sharp, the height of the peak can be used instead of the area under the peak.

b   The retention time of A = 2.0 min; B = 8.8 min; C = 19.0 min

c   Substance C has the greatest retention time, so it spends more time in the column and consequently has least affinity for the mobile phase.

Summary questions

1   Often the outlet from a GLC instrument is connected to a mass spectrometer. Why is this beneficial? *(1 mark)*

2   What would be the effect on the retention time of increasing the flow rate of the carrier gas? *(1 mark)*

3   Sketch the output you would expect from a GLC instrument for a mixture containing compounds X, Y, and Z in the table below. *(1 mark)*

Compound	Retention time
X	3 min
Y	5.5 min
Z	6.4 min

Chapter 7 Experimental techniques

Determining a solubility product, K_{sp}

The first stage in determining K_{sp} is to make a saturated solution of the salt in question, in distilled water.

1 Warm distilled water in a conical flask and add the salt. Keep adding the salt until no more dissolves. Allow the mixture to cool to room temperature, then filter the mixture, discarding the residue.

   Next, the concentration of one of the ions in solution must be determined. This allows K_{sp} to be determined. It is only necessary to determine the concentration of one of the ions as the concentration of the other ion will be proportional to it.

2 Take the temperature of the solution, as K_{sp} is temperature-dependent.

3 Decide the appropriate method for determining the concentration of the ion:

   a for a basic solution such as calcium hydroxide, the concentration of hydroxide ions can be determined using a titration with hydrochloric acid.

   b for a coloured solution such as one containing a transition metal ion, the concentration of the coloured ion can be determined using colorimetry.

Determining an equilibrium constant

1 Allow the mixture of reactants to reach equilibrium.

2 Record the temperature, as K_c is temperature-dependent.

3 Determine the concentration of one of the components. The method will depend on the substances involved, but could include titrations, colorimetry, or pH measurements.

4 The concentrations of the other components can be calculated from the concentration determined in step 3, using the chemical equation for the reaction. You will also need to know the initial concentrations of the substances.

5 Write an expression for K_c and substitute the values you have calculated. Determine the value of K_c including its units. Report the temperature.

> **Revision tip**
>
> Tap water must not be used as it contains dissolved ions which would affect the calculation.

> **Revision tip**
>
> For example, in AgCl, the ratio is 1:1, and in $Ca(OH)_2$, the ratio is 1:2.

1 What is the only factor that affects the magnitude of K_c?

   A Concentration

   B Pressure

   C Surface area

   D Temperature   *(1 mark)*

2 Catalysts do **not**…

   A Speed up a reaction

   B Provide an alternative route of lower activation enthalpy

   C Alter the position of equilibrium

   D Remain unchanged at the end of the reaction   *(1 mark)*

3 What is the effect on K_c of increasing the temperature of an exothermic reaction?

   A K_c is unchanged

   B The position of equilibrium moves to the right

   C The value of K_c decreases

   D Collisions occur more frequently   *(1 mark)*

4 What is the correct expression for the solubility product of calcium carbonate?

   A $K_{sp} = \dfrac{[Ca^{2+}][CO_3^{2-}]}{[CaCO_3]}$

   B $K_{sp} = \dfrac{[CaCO_3]}{[Ca^{2+}][CO_3^{2-}]}$

   C $K_{sp} = [Ca^{2+}][CO_3^{2-}][CaCO_3]$

   D $K_{sp} = [Ca^{2+}][CO_3^{2-}]$ *(1 mark)*

5 **a** Write an expression for K_c for the reaction: $C_2H_5OH \rightleftharpoons C_2H_4 + H_2O$ *(1 mark)*

   b Deduce the units of K_c for this reaction.   *(1 mark)*

6 The following equilibrium takes place during the production of sulfuric acid:

$$2SO_2 + O_2 \rightleftharpoons 2SO_3 \qquad \Delta H = -196\,kJ\,mol^{-1}$$

   a Write an expression for K_c for the reaction.   *(1 mark)*

   b Explain the effect on the position of equilibrium and the magnitude of K_c of:
(i) increasing the pressure, (ii) adding a catalyst, (iii) increasing the temperature.   *(3 marks)*

   c Calculate the value of K_c, including units, when $[SO_2] = 7.45 \times 10^{-3}\,mol\,dm^{-3}$, $[O_2] = 3.62 \times 10^{-3}\,mol\,dm^{-3}$ and $[SO_3] = 1.80 \times 10^{-2}\,mol\,dm^{-3}$.   *(2 marks)*

7 **a** Write an expression for K_{sp} for lead(II) sulfide, PbS.   *(1 mark)*

   b If the concentration of Pb^{2+} ions is $1.14 \times 10^{-14}\,mol\,dm^{-3}$, what is the maximum concentration of S^{2-} ions that will not cause a precipitate to form? K_{sp} (PbS) $= 1.3 \times 10^{-28}\,mol^2\,dm^{-6}$.   *(2 marks)*

8 A 3:1 mixture of octane and decane was separated by gas chromatography. The retention time of octane was 5.5 minutes and the retention time of decane was 7.5 minutes.

   a Explain what is meant by 'retention time'.   *(1 mark)*

   b Describe how the areas under each peak will compare to each other.   *(1 mark)*

   c Explain how the identity of each component could be confirmed by mass spectrometry.   *(1 mark)*

8.2 Strong and weak acids and pH

Specification reference: O(i), O(j), O(k), O(l), O(n)

Acids and bases

One way of characterising acids and bases is through their ability to transfer or accept H^+ ions. This is the Brønsted–Lowry theory.

Proton transfer

An acid is a proton (H^+) donor. In aqueous solution, an acid donates protons to water molecules to form oxonium ions (H_3O^+). These are often abbreviated to $H^+(aq)$.

Acid–base pairs

In many cases, the donation of a proton by an acid is reversible:

$$HA\ (aq) \rightleftharpoons H^+(aq) + A^-(aq)$$

HA donates protons and acts as an acid. In the reverse direction, A^- acts as a base as it gains a proton to reform HA. So HA and A^- are a **conjugate acid–base pair**.

Strength of acids and bases

Acids vary in their ability to donate protons. Acids which are powerful proton (H^+) donors are called **strong acids**. **Weak acids** are moderate or poor proton (H^+) donors. A strong acid has a weak conjugate base, and vice versa.

In strong acids, almost all of the acid molecules donate their protons – the acid undergoes **complete dissociation**. Examples of strong acids are hydrochloric acid (HCl), sulfuric acid (H_2SO_4), and nitric acid (HNO_3).

In weak acids, only a small proportion of the acid molecules donate their protons – the acid undergoes **incomplete dissociation**. Carboxylic acids such as ethanoic acid are weak acids:

$$CH_3COOH(aq) \rightleftharpoons CH_3COO^-(aq) + H^+(aq)$$

The acidity constant K_a

K_a is **the acidity constant** or **acid dissociation constant**. The greater the value of K_a, the stronger the acid:

$$HA(aq) \rightleftharpoons A^-(aq) + H^+(aq)$$

$$K_a = \frac{[A^-][H^+]}{[HA]}$$

When comparing weak acids, which can have *very* small K_a values, the K_a can be converted into a pK_a value for ease of use:

$$pK_a = -\log K_a$$

Calculating pH

$$pH = -\log[H^+(aq)]$$

Strong acids

Because a strong acid is fully dissociated, we can assume that the concentration of acid in the solution is the same as the concentration of the hydrogen ions in that solution if the acid is monoprotic.

Revision tip

Remember H^+ is a proton. An acid is a proton donor and a base is a proton acceptor.

Key term

The conjugate acid: This is so called because in the reverse reaction it is the species that donates protons.

Revision tip

Just as all acids have a conjugate base, so all bases have a conjugate acid.

Key terms

Strong acid: An acid which is fully ionised into H^+ and A^- ions.

Weak acid: An acid which has undergone incomplete ionisation into H^+ and A^- ions.

Revision tip

Sometimes the word 'dissociation' is used instead of ionisation when referring to acids.

Common misconception: Weak acids

Equations showing dissociation of weak acids must contain an equilibrium arrow $\rightleftharpoons$; strong acids have a normal arrow $\rightarrow$ as the dissociation is complete.

Revision tip

Remember that square brackets around a formula means the concentration of whatever is inside the brackets, in $mol\,dm^{-3}$.

Revision tip

pH has no units.

Revision tip

For dilute solutions the normal range is 0–14. Acids range from pH 0–6, and alkalis range from pH 8–14.

Revision tip

Monoprotic acids have one acidic hydrogen – they release one proton per molecule.

Revision tip

Assumption 1 is more reliable for very weak acids, where K_a is small. If the weak acid has significant dissociation, assumption 1 can be a source of inaccuracy in the calculation of pH.

Revision tip

The temperature should always be quoted because K_a, and ultimately pH, varies with temperature.

Revision tip

K_w is the symbol used to represent the ionic product for water.

 Worked example: The pH of a strong acid

Calculate the pH of a 0.001 mol dm⁻³ solution of hydrochloric acid.

Step 1: $HCl(aq) \rightarrow H^+(aq) + Cl^-(aq)$

Step 2: $[H^+] = 0.001 \text{ mol dm}^{-3}$

Step 3: $pH = -\log 0.001 = 3.00$

 Worked example: The pH of a diprotic acid

Calculate the pH of a 0.005 mol dm⁻³ solution of sulfuric acid.

Step 1: $H_2SO_4(aq) \rightarrow 2H^+(aq) + SO_4^{2-}(aq)$

Step 2: $[H^+] = 0.005 \text{ mol dm}^{-3} \times 2 = 0.01 \text{ mol dm}^{-3}$

Step 3: $pH = -\log 0.01 = 2.00$

The concentration of H^+ is twice that of the H_2SO_4 because 1 mole of the acid dissociates to produce 2 moles of hydrogen ions.

Weak acids

Two assumptions are made:

- Assumption 1: the equilibrium concentration [HA] for a weak acid is the same as the initial concentration of the acid.
- Assumption 2: The equilibrium concentration [A⁻] is equal to the equilibrium concentration [H⁺]. A few protons will be provided by the water, but these are insignificant compared to those provided by the acid.

 Worked example: The pH of a weak acid

Calculate the pH of 0.01 mol dm⁻³ CH_3COOH ($K_a = 1.7 \times 10^{-5}$ mol dm⁻³ at 298 K).

$$CH_3COOH(aq) \rightleftharpoons CH_3COO^-(aq) + H^+(aq)$$

Step 1: Write an expression for K_a: $K_a = \dfrac{[CH_3COO^-][H^+]}{[CH_3COOH]}$

Step 2: Assumption 1: We assume few molecules dissociate, so $[CH_3COOH(aq)] = 0.01 \text{ mol dm}^{-3}$.

Step 3: Assumption 2: We can assume for every 1 mole of H^+ present there is 1 mole of CH_3COO^-, so $[H^+(aq)] = [CH_3COO^-(aq)]$.

Step 4: $K_a = \dfrac{[H^+]^2}{0.01} = 1.7 \times 10^{-5}$

Step 5: $[H^+] = \sqrt{1.7 \times 10^{-7}}$

Step 6: $[H^+] = 4.12 \times 10^{-5}$

Step 7: $pH = -\log_{10}(4.12 \times 10^{-5}) = 3.4$

Strong bases

For a strong base, we can assume that $[OH^-(aq)]$ is equal to the concentration of the solution of the base, because it is fully dissociated in aqueous solution. To then calculate $[H^+(aq)]$, the ionic product of water is used.

Water dissociates slightly and this gives rise to the equilibrium constant K_w:

$$H_2O(l) \rightleftharpoons H^+(aq) + OH^-(aq)$$

$$K_w = [H^+(aq)][OH^-(aq)] = 1 \times 10^{-14} \text{ mol}^2 \text{ dm}^{-6} \text{ at 298 K}$$

Worked example: The pH of a strong base

Calculate the pH of a 0.01 mol dm^{-3} solution of NaOH.

Step 1: $NaOH(aq) \rightarrow Na^+(aq) + OH^-(aq)$

Step 2: The strong base is fully dissociated, so $[OH^-(aq)] = 0.01$ mol dm^{-3}.

Step 3: $K_w = [H^+(aq)][OH^-(aq)]$

$$[H^+] = \frac{K_w}{[OH^-(aq)]} = \frac{1 \times 10^{-14} \, mol^2 \, dm^{-6}}{0.01 \, mol \, dm^{-3}} = 1 \times 10^{-12} \, mol \, dm^{-3}$$

Step 4: $pH = -\log(1 \times 10^{-12}) = 12.0$

The greenhouse effect

The earth is kept warm by the greenhouse effect, which is a natural phenomenon. However increased concentrations of CO_2, principally from burning fossil fuels, have led to an enhanced greenhouse effect. Solar energy reaches Earth mainly as visible and ultraviolet radiation. The Earth absorbs some of this energy and radiates some in the form of infrared radiation. Water vapour in the atmosphere absorbs certain wavelengths of infrared, but greenhouse gases such as carbon dioxide and methane in the troposphere absorb the re-radiated infrared in the 'IR window'. The 'IR window' refers to the wavelengths that are not absorbed by water molecules. The absorption of infrared by greenhouse gas molecules increases the vibrational energy of their bonds. This energy is transferred to other molecules by collisions, which increases their kinetic energy and raises the temperature. Greenhouse gas molecules also re-emit some of the absorbed infrared in all directions, which contributes to the heating of the Earth.

The oceans can absorb carbon dioxide and prevent it building up in the atmosphere. The carbon dioxide can react with the water, acting as a base as it accepts H^+ ions to form hydrogencarbonate ions (see Reaction 1). Some hydrogencarbonate ions dissociate to form carbonate ions (see Reaction 2):

Reaction 1 $CO_2(aq) + H_2O(l) \rightleftharpoons HCO_3^-(aq) + H^+(aq)$

Reaction 2 $HCO_3^-(aq) \rightleftharpoons CO_3^{2-}(aq) + H^+(aq)$

However, these reactions are responsible for acidification of the oceans as H^+ ions are formed.

Summary questions

1 In the reaction below, identify the two conjugate acid–base pairs:
   $H_2SO_4 + OH^- \rightarrow HSO_4^- + H_2O$ *(2 marks)*

2 Calculate the pH of 0.10 mol dm^{-3} nitric acid.   *(1 mark)*

3 Calculate the pH of 0.05 mol dm^{-3} potassium hydroxide solution.   *(2 marks)*

4 **a** Give the expression for the K_a of methanoic acid (HCOOH).   *(1 mark)*
   b Calculate the pH of 0.001 mol dm^{-3} methanoic acid
       $(K_a = 1.60 \times 10^{-4} \, mol \, dm^{-3})$.   *(2 marks)*

8.3 Buffer solutions

Specification reference: O(m)

Key term

Buffer: A solution which minimises the change in pH on addition of small quantities of acid or alkali.

Synoptic link

You learned about conjugate acid–base pairs in Topic 8.2, Strong and weak acids and pH.

Synoptic link

You learned about equilibria in Topic 7.1, Chemical equilibrium.

What is a buffer solution?

Buffer solutions are solutions that have an almost constant pH, despite dilution or small additions of acid or alkali. Buffer solutions contain either:

- a weak acid and one of its salts, e.g. ethanoic acid and sodium ethanoate, or
- a weak base and one of its salts, e.g. ammonia and ammonium chloride.

All buffer solutions contain large amounts of a proton donor – a weak acid or conjugate acid – and large amounts of a proton acceptor, in other words a weak base or conjugate base. Any additions of acid or alkali react with these large amounts, and this keeps the pH constant within limits.

Take the buffer system comprising ethanoic acid and sodium ethanoate solution as an example. The weak acid (ethanoic) partially dissociates to produce its conjugate base and protons. In this case, the acid is a weak acid, so the position of equilibrium lies to the left:

$$CH_3COOH(aq) \rightleftharpoons CH_3COO^-(aq) + H^+(aq) \qquad \textbf{Equation 1}$$
$$\text{weak acid} \qquad\qquad \text{conjugate base}$$

If small amounts of alkali are added, the weak acid dissociates to produce more H^+ ions, which react with the added OH^- ions: the position of equilibrium in Equation 1 moves to the right, so maintaining the pH.

If small amounts of acid are added, the added H^+ ions react with the ethanoate ions present. However, it is not long before the ethanoate ions have all reacted. For the buffer to successfully resist a change in pH, larger quantities of ethanoate ions are required, which is why sodium ethanoate was added when the buffer is produced. Then the added H^+ ions can react with the large amounts of ethanoate ions from the salt – the position of the equilibrium in Equation 1 moves to the left, so maintaining the pH. In summary:

- H^+ ions are added.
- The position of equilibrium moves to the left.
- pH is maintained.
- There is a large concentration of the salt.

Calculations with buffers

Recall the K_a expression:

$$K_a = \frac{[A^-][H^+]}{[HA]}$$

We make two assumptions in buffer calculations:

- All the anions $[A^-]$ have come from the salt, so the contribution from the acid is negligible, i.e. $[A^-]$ = [salt].
- The concentration of the acid in solution [HA(aq)] is the same as the amount of acid put into the solution, in other words, ignore any dissociation, i.e. [HA] = [acid].

Using these assumptions, it can be shown that $K_a = [H^+]\dfrac{[salt]}{[acid]}$

Finding the pH of a buffer solution

If K_a of the weak acid is known, along with the concentrations of salt and weak acid, then the hydrogen ion concentration can be calculated, and hence the pH.

Revision tip

Remember $pH = -\log_{10}[H^+]$

> **▣ Worked example: Calculating the pH of a buffer**
>
> Calculate the pH of a buffer solution made by mixing equal volumes of $0.20 \, mol \, dm^{-3}$ ethanoic acid and $0.10 \, mol \, dm^{-3}$ sodium ethanoate solutions (for ethanoic acid, $K_a = 1.7 \times 10^{-5} \, mol \, dm^{-3}$ at 298 K).
>
> **Step 1:** By mixing equal quantities, each original concentration will be halved, so:
>
> $$[CH_3COOH(aq)] = 0.10 \, mol \, dm^{-3} \text{ and } [CH_3COO^-(aq)] = 0.05 \, mol \, dm^{-3}$$
>
> **Step 2:** $K_a = [H^+]\dfrac{[salt]}{[acid]}$
>
> **Step 3:** $[H^+] = K_a\dfrac{[acid]}{[salt]} = \dfrac{1.7 \times 10^{-5} \, mol \, dm^{-3} \times 0.10 \, mol \, dm^{-3}}{0.05 \, mol \, dm^{-3}} = 3.4 \times 10^{-5} \, mol \, dm^{-3}$
>
> **Step 4:** $pH = -\log_{10}(3.4 \times 10^{-5}) = 4.5$

Buffers in action

Buffers are found in shampoos, and in food and drink where they are often referred to as 'acidity regulators'. Buffers in the blood protect us from changes in pH due to formation of CO_2 and H^+ in metabolic processes. Otherwise these pH changes could affect the action of enzymes, and have serious consequences for your health.

> **Summary questions**
>
> 1 Calculate the pH of a solution containing equal amounts of benzoic acid and sodium benzoate, where the ratio $\dfrac{[acid]}{[salt]} = 1$. $K_a = 6.3 \times 10^{-5} \, mol \, dm^{-3}$ *(2 marks)*
>
> 2 Calculate the pH of a buffer in which $[CH_3COOH] = 0.001 \, mol \, dm^{-3}$ and $[CH_3COONa] = 0.005 \, mol \, dm^{-3}$. $K_a = 1.7 \times 10^{-5} \, mol \, dm^{-3}$ *(2 marks)*
>
> 3 Calculate the pH of a solution made by mixing equal volumes of $0.02 \, mol \, dm^{-3}$ methanoic acid and $0.012 \, mol \, dm^{-3}$ potassium methanoate solution. $K_a = 1.8 \times 10^{-4} \, mol \, dm^{-3}$ *(2 marks)*

Measuring pH

A pH electrode is used to measure pH accurately. It must be calibrated first using solutions of known pH.

1 Wash the electrode with distilled water, then transfer it to a buffer solution of pH 7.00. Check the bulb is completely immersed and, once the reading is stable, ensure it reads 7.00. Adjust if necessary.

2 If you wish to measure the pH of acidic solutions, further calibrate the electrode with an acidic buffer solution, such as a pH 4.00 buffer.

3 If you wish to measure the pH of alkaline solutions, further calibrate the electrode with an alkaline buffer solution, such as a pH 10.00 buffer.

4 If you wish to measure the pH of both acidic and alkaline solutions, calibrate the electrode with acidic and alkaline buffer solutions.

5 The pH electrode can then be used to measure the pH of the test solution, by immersing it in the solution to be measured.

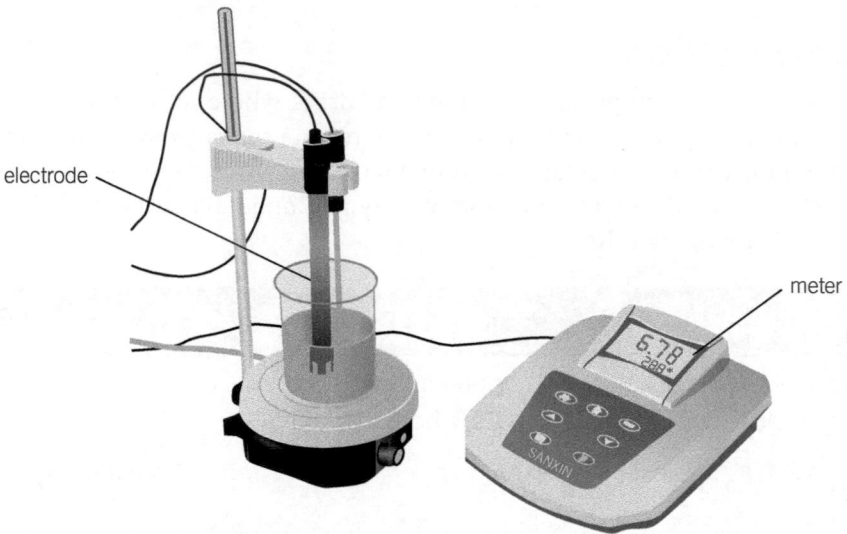

electrode

meter

▲ **Figure 1** *Measurement of pH*

Chapter 8 Practice questions

1 A buffer solution…

    A Consists of a strong acid and its salt

    B Consists of a strong base and its salt

    C Is significantly affected by additions of acid or alkali

    D Contains approximately equal concentrations of acid (or base)
       and salt   *(1 mark)*

2 Which statement is correct?

    A The pH of a strong acid is equal to the concentration of the acid

    B $[H^+]$ for a strong acid is equal to the concentration of the acid

    C $[H^+]$ for a strong alkali is equal to $K_a \times \dfrac{[\text{acid}]}{[\text{salt}]}$

    D The pH of a weak acid is equal to $-\log_{10}[HA]$ *(1 mark)*

3 Look at the following statements about the reaction:

    $NH_3 + H_2O \rightleftharpoons NH_4^+ + OH^-$

    1 NH_3 is acting as a base

    2 H_2O is acting as an acid

    3 OH^- is a conjugate acid

    Which statements are correct?

    A 1 only

    B 1 and 2 only

    C 2 and 3 only

    D 1, 2, and 3   *(1 mark)*

4 What is the pH of $0.05 \, \text{mol dm}^{-3} \, H_3PO_4$?

    A −1.30

    B 0.05

    C 1.30

    D 0.82   *(1 mark)*

5 What is the pH of $0.05 \, \text{mol dm}^{-3}$ propanoic acid $(K_a = 1.3 \times 10^{-5} \, \text{mol dm}^{-3})$?

    A 8.06×10^{-4}

    B 0.05

    C 2.44

    D 3.09   *(1 mark)*

6 Explain why a solution of a weak acid on its own *does not* act as a
buffer solution.   *(1 mark)*

7 A student prepares a sample of $0.01 \, \text{mol dm}^{-3}$ benzoic acid, then adds an
equal volume of $0.01 \, \text{mol dm}^{-3}$ sodium benzoate. Calculate the pH of this
buffer and explain how it minimises changes to pH.
K_a (benzoic acid) $= 6.3 \times 10^{-5} \, \text{mol dm}^{-3}$.   *(5 marks)*

9.3 Redox reactions, cells, and electrode potentials

Specification reference: DM(c), DM(d)

Synoptic link

Oxidation and reduction are introduced in Topic 9.1, Oxidation and reduction.

Key term

Redox reaction: A reaction involving simultaneous oxidation and reduction.

Redox reactions

You have seen that oxidation is the loss of electrons and reduction is gain of electrons. A redox reaction is a reaction in which oxidation and reduction occur at the same time. The overall equation for the reaction can be split into two parts, called half-equations, which show the species that gain or lose electrons. Half-equations involve ions and electrons.

Combining half-equations

Adding together the two half-equations gives the overall equation for the redox reaction.

Worked example: Writing redox equations

Write the combined equation for the displacement reaction when chlorine reacts with bromide ions.

Step 1: Write the half-equations for the oxidation and reduction reactions:

$$2Br^- \rightleftharpoons Br_2 + 2e^- \quad \textbf{oxidation half-equation}$$

$$Cl_2 + 2e^- \rightleftharpoons 2Cl^- \quad \textbf{reduction half-equation}$$

Step 2: Make sure the number of electrons is the same in each half-equation. Here there are two electrons in each half-equation.

Step 3: Add the two half-equations together, cancelling electrons from the left- and right-hand sides.

$$2Br^- + Cl_2 \rightarrow Br_2 + 2Cl^-$$

Model answer: Writing redox equations

Write the combined equation for the reaction of chromate(VI) ions, $Cr_2O_7^{2-}$, with sulfate(IV) ions, SO_3^{2-}.

Step 1: Write the half-equations for the oxidation and reduction reactions:

$$SO_3^{2-}(aq) + H_2O(l) \rightleftharpoons SO_4^{2-}(aq) + 2H^+(aq) + 2e^- \quad \textbf{oxidation half-reaction}$$

$$Cr_2O_7^{2-}(aq) + 14H^+(aq) + 6e^- \rightleftharpoons 2Cr^{3+}(aq) + 7H_2O(l) \quad \textbf{reduction half-reaction}$$

Step 2: Make sure the number of electrons is the same in each half-equation. In this case, there needs to be 6 electrons in each half-equation:

> The oxidation half-equation must be multiplied by a factor of 3 to produce 6 electrons, which are accepted by the reduction half-equation.

$$3SO_3^{2-}(aq) + 3H_2O(l) \rightleftharpoons 3SO_4^{2-}(aq) + \textbf{6}H^+(aq) + \textbf{6}e^-$$

$$Cr_2O_7^{2-}(aq) + 14H^+(aq) + \textbf{6}e^- \rightleftharpoons 2Cr^{3+}(aq) + 7H_2O(l)$$

Step 3: Add the two half-equations together, cancelling electrons and hydrogen ions from the left- and right-hand sides:

$$Cr_2O_7^{2-}(aq) + 8H^+(aq) + 3SO_3^{2-}(aq) \rightarrow 2Cr^{3+}(aq) + 4H_2O(l) + 3SO_4^{2-}(aq)$$

Revision tip

Electrode potentials allow you to decide which is the oxidation half-reaction and which is the reduction half-reaction.

Electrode potentials

If you place a strip of metal in an aqueous solution of its ions, an electrode potential, or potential difference, is created. An equilibrium is established between the metal and the ions. For example:

$$Cu^{2+}(aq) + 2e^- \rightleftharpoons Cu(s)$$

$$Zn^{2+}(aq) + 2e^- \rightleftharpoons Zn(s)$$

Like all equilibria, these are affected by temperature and the concentration of the ions, and the conditions should always be stated.

Metals differ in their tendency to release electrons. Metals which release electrons more readily have a more negative electrode potential. Altering the temperature or the concentration of ions in solution alters the value of the electrode potential.

Electrochemical cells

An electrochemical cell consists of two half-cells joined together, with a high-resistance voltmeter measuring the maximum potential difference between two half-cells. To complete the circuit a salt bridge provides an ionic connection between two half-cells. It is often made from a strip of filter paper soaked in a saturated solution of potassium nitrate.

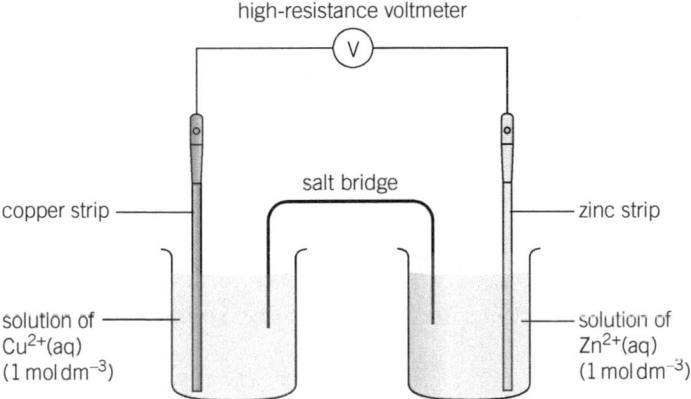

▲ **Figure 1** *A copper–zinc cell*

It is not possible to measure the electrode potential of a single half-cell. To measure the electrode potential of a metal/ion system, the two half-cells must be joined together to form an electrochemical cell.

The potential difference between the two half-cells is called the cell potential, E_{cell}. Electrons flow through the wires from the negative terminal to the positive terminal:

- The half-cell with the more negative electrode potential forms the negative terminal of the cell.
- The half-cell with the more positive electrode potential forms the positive terminal of the cell.

Standard electrode potentials

It is necessary to have a standard set-up to compare the electrode potentials of all metal/ion half-cells. To do this a standard hydrogen half-cell is used. Its electrode potential under standard conditions is defined as 0.00 V.

The half-reaction occurring in the standard hydrogen half-cell is:

$$H^+(aq) + e^- \rightleftharpoons \frac{1}{2}H_2(g)$$

The **standard electrode potential**, $E^\ominus$, of a half-cell is the potential difference between the half-cell and a standard hydrogen half-cell.

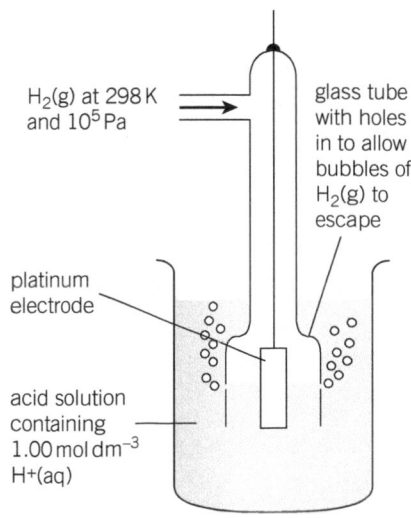

▲ **Figure 2** *The standard hydrogen half-cell*

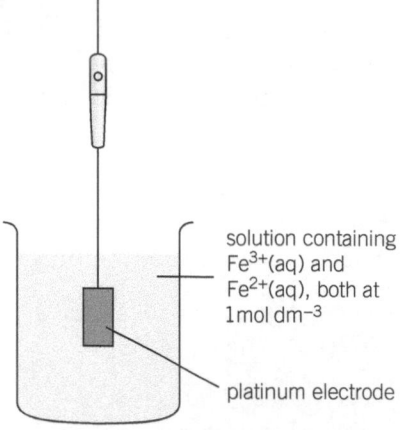

solution containing $Fe^{3+}(aq)$ and $Fe^{2+}(aq)$, both at $1\,mol\,dm^{-3}$

platinum electrode

▲ **Figure 3** *A standard half-cell for the* $Fe^{3+}(aq)/Fe^{2+}(aq)$ *half-reaction*

Revision tip

The half-equations for $I_2/2I^-$ and Fe^{3+}/Fe^{2+} are:

$$I_2 + 2e^- \rightleftharpoons 2I^-$$

$$Fe^{3+} + e^- \rightleftharpoons Fe^{2+}$$

Revision tip

Remember that all concentrations in half-cells are $1.00\,mol\,dm^{-3}$.

Revision tip

Both half-cells might have negative $E^{\varnothing}$ values. In this case, the one with the least negative value has the most positive $E^{\varnothing}$.

Revision tip

$E^{\varnothing}$ cell values do not have a sign as they represent the *difference* in potential.

Revision tip

Oxidation is the loss of electrons, which occurs at the negative electrode. Reduction is the gain of electrons, which occurs at the positive electrode.

To measure a standard electrode potential, the half-cell being investigated is connected to a standard hydrogen half-cell.

For half-cells such as $I_2/2I^-$ or Fe^{3+}/Fe^{2+}, which do not feature a metal in elemental form, an inert electrode such as platinum is dipped into a solution containing all the ions and molecules involved in the half-reaction.

🖩 **Worked example: Calculating $E^{\varnothing}_{cell}$**

What is $E^{\varnothing}_{cell}$ when the Fe^{2+}/Fe and Cu^{2+}/Cu half-cells are connected?

Step 1: Look up the standard electrode potentials for the two half-reactions:

$$Fe^{2+}(aq) + 2e^- \rightleftharpoons Fe(s) \qquad E^{\varnothing} = -0.44\,V$$

$$Cu^{2+}(aq) + 2e^- \rightleftharpoons Cu(s) \qquad E^{\varnothing} = +0.34\,V$$

Step 2: Calculate $E^{\varnothing}_{cell} = E^{\varnothing}$ [most positive electrode] $- E^{\varnothing}$ [most negative electrode]

$$E^{\varnothing}_{cell} = +0.34\,V - (-0.44\,V) = 0.78\,V$$

Predicting the direction of a reaction

Electrons are released from the half-cell with the more negative electrode potential, where oxidation occurs. Electrons are accepted by the half-cell with the more positive electrode potential, where reduction occurs.

It is possible to use this idea to predict the feasibility of a reaction, by calculating $E^{\varnothing}_{cell}$ for the proposed reaction.

Model answer: Predicting the feasibility of a reaction

Explain whether the reaction between aqueous chlorine and aqueous iodide ions is feasible.

Step 1: Look up the half-reactions and their standard electrode potentials:

$$I_2(aq) + 2e^- \rightleftharpoons 2I^-(aq) \qquad E^{\varnothing} = +0.54\,V \qquad \text{half-reaction 1}$$

$$Cl_2(g) + 2e^- \rightleftharpoons 2Cl^-(aq) \qquad E^{\varnothing} = +1.36\,V \qquad \text{half-reaction 2}$$

Step 2: Construct an electrode potential chart.

> Although both half-cells have a positive $E^{\varnothing}$, the I_2/I^- is the more negative half-cell.

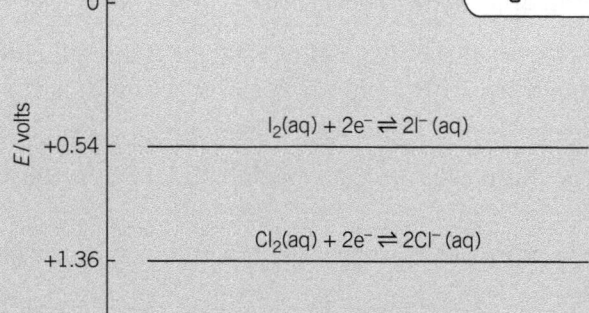

Step 3: Use the electrode potential chart to predict whether a reaction could occur.

Electrons flow to the positive terminal, which is the $Cl_2(aq)/Cl^-(aq)$ half-cell, which has an electrode potential of $+1.36\,V$.

A reduction reaction will occur in the positive half-cell:

$$Cl_2(g) + 2e^- \rightleftharpoons 2Cl^-(aq)$$

Step 4: The other half-cell must supply electrons and is the negative terminal of the cell. Oxidation occurs.

$$2I^-(aq) \rightleftharpoons I_2(aq) + 2e^-$$

Step 5: Use the half-equations to give an overall equation.

$$2I^-(aq) + Cl_2(g) \rightarrow I_2(aq) + 2Cl^-(aq)$$

Therefore the reaction that occurs is that of iodide ions with aqueous chlorine. The reverse reaction does not occur, so chloride ions *do not* react with aqueous iodine.

Remember that this method predicts the feasibility of a reaction occurring under standard conditions. If the activation enthalpy is high, the reaction may not actually happen. However a catalyst might enable the reaction to occur.

Revision tip

Changing the conditions such as concentration and/or temperature will change the $E^\varnothing$ cell values, which may also cause the reaction to happen.

Summary questions

1   Describe the experimental set-up and conditions required to determine $E^\varnothing_{cell}$ for a copper–silver cell. (*5 marks*)

2   Write a balanced equation for the reaction of manganate(VII) ions with iron(II) ions. The half-equations are:
$MnO_4^-(aq) + 8H^+(aq) + 5e \rightleftharpoons Mn^{2+}(aq) + 4H_2O(aq)$
and $Fe^{3+}(aq) + e^- \rightleftharpoons Fe^{2+}(aq)$ (*2 marks*)

3   Calculate $E^\varnothing_{cell}$ for the reaction in question 2, and explain why Mn^{2+} ions do not reduce Fe^{3+} ions. $E^\varnothing(MnO_4^-/Mn^{2+}) = +1.51$ V; $E^\varnothing(Fe^{3+}/Fe^{2+}) = +0.77$ V (*2 marks*)

9.4 Rusting and its prevention

Specification reference: DM(f)

Rusting

Corrosion of iron, commonly known as rusting, is a widespread problem, and it can have serious economic consequences. It occurs due to electrochemical processes, and certain methods of rust prevention also rely on electrochemistry.

Half-equations in rusting

Rusting occurs in a series of steps. The half-equations involved are:

$$Fe^{2+}(aq) + 2e^- \rightleftharpoons Fe(s) \qquad E^{\varnothing} = -0.44\,V$$

$$O_2(g) + 2H_2O(l) + 4e^- \rightleftharpoons 4OH^-(aq) \qquad E^{\varnothing} = +0.40\,V$$

$$E^{\varnothing}_{cell} = +0.84\,V$$

The Fe/Fe^{2+} half-cell is more negative, so it releases electrons. This means that Fe atoms are oxidised into Fe^{2+} ions. The electrons flow to the half-cell comprising oxygen dissolved in a water droplet, which accepts the electrons and produces hydroxide ions.

Two further reactions then occur. First the Fe^{2+} and OH^- ions react to form solid iron(II) hydroxide:

$$Fe^{2+}(aq) + 2OH^-(aq) \rightarrow Fe(OH)_2(s)$$

> **Revision tip**
>
> This is an example of a precipitation reaction.

Secondly the iron(II) hydroxide reacts with further oxygen to form hydrated iron(III) oxide, $Fe_2O_3 \cdot xH_2O(s)$, which is rust.

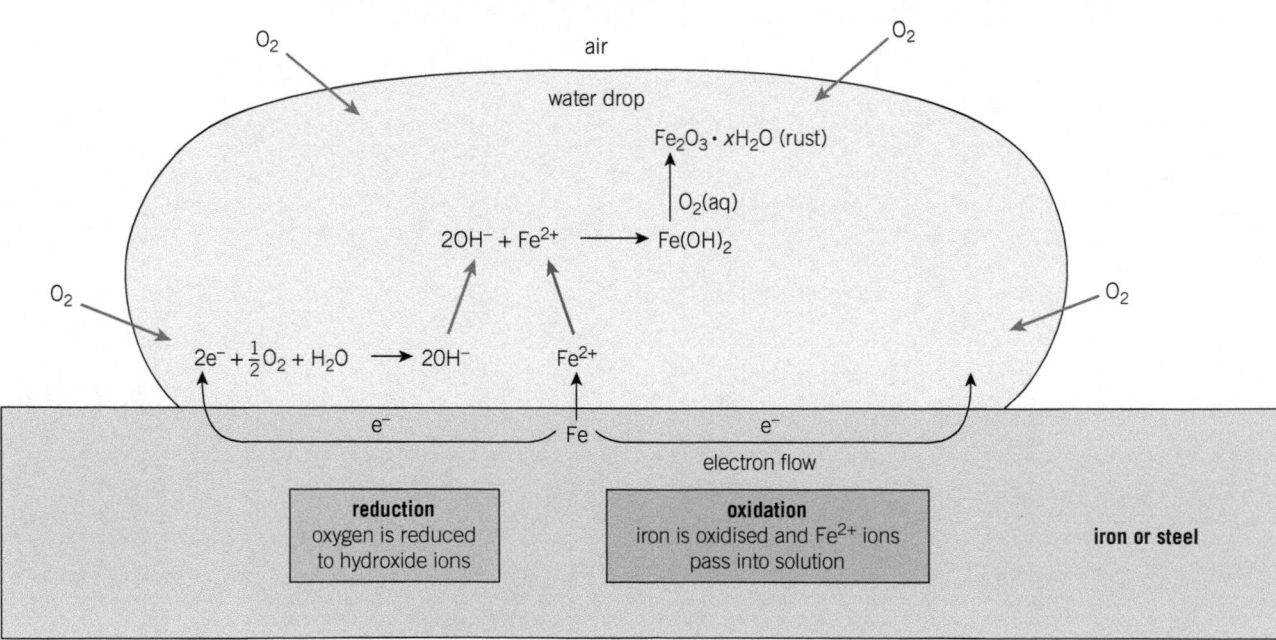

▲ **Figure 1** *Rusting is an electrochemical process*

Preventing rusting

A common means of protecting iron objects from rusting is to paint, grease, or oil them. This provides a barrier between the metal and atmospheric oxygen.

The iron can be covered with a layer of another metal, such as zinc. This is known as **galvanising**. Although the zinc reacts with oxygen, it produces a hard layer of zinc oxide which ultimately protects the iron underneath.

If the zinc coating is damaged, it will still protect against rusting. This is because $E^{\ominus}$ for the zinc half-cell is more negative than $E^{\ominus}$ for the iron half-cell:

$$Zn^{2+}/Zn \quad E^{\ominus} = -0.76\,V$$

$$Fe^{2+}/Fe \quad E^{\ominus} = -0.44\,V$$

As a result, the zinc corrodes in preference to the iron. This is known as **sacrificial protection**. Large iron objects are sometimes protected by having blocks of metals such as zinc or magnesium, which have more negative electrode potentials, applied to them.

Summary questions

1 Explain why rusting only occurs in the presence of oxygen and water.
(*1 mark*)

2 Identify which species are oxidised and reduced in the half-equations for rusting, by stating the change in oxidation number. (*2 marks*)

3 Calculate $E^{\ominus}_{cell}$ for the reaction of zinc with the O_2/OH^- half-cell, and suggest why zinc corrodes before iron. (*2 marks*)

Key term

Sacrificial protection: The use of a metal with a more negative electrode potential to prevent rusting. The other metal corrodes first.

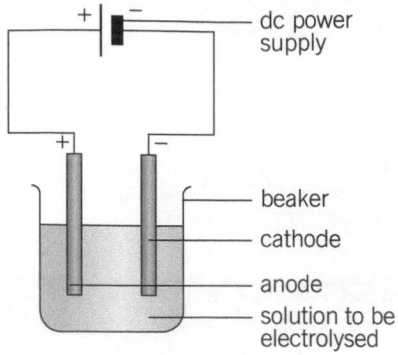

▲ **Figure 1** *Apparatus for electrolysis*

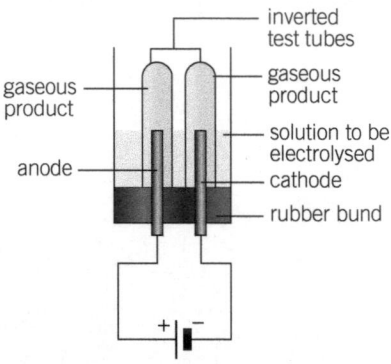

▲ **Figure 2** *Apparatus to collect gaseous products in electrolysis*

Revision tip

Standard temperature and concentration for electrochemical cells is 298 K and 1.0 mol dm⁻³.

Electrolysis of aqueous solutions

An electric current is passed through an electrolyte. Different set-ups are required, depending whether the products of electrolysis are solids or gaseous.

The test tubes are filled with water at the start of the reaction, and the gaseous products displace the water.

Electrolysis can be used to purify a metal, in which case the anode is made of the impure metal. The electrolyte must contain ions of the metal and the cathode must be made of the pure metal.

Electrochemical cells

The potential of an electrochemical cell can be determined by connecting two half-cells. Standard electrode potentials are measured by connecting the half-cell in question to a standard hydrogen half-cell, or a calibrated reference half-cell.

1 Start by constructing the half-cell of interest:

● For a metal/metal ion half-cell such as Cu^{2+}/Cu, the electrode is a strip of metal dipping into a 1.0 mol dm⁻³ solution of the metal ion.

● For a half-cell containing two ions of the same element such as Fe^{3+}/Fe^{2+}, the electrode is a platinum (or graphite) rod dipping into a mixture of the two ions. The concentration must be 1.0 mol dm⁻³.

● The temperature must be 298 K.

2 Connect the half-cell to the reference cell using a salt bridge and a high-resistance voltmeter.

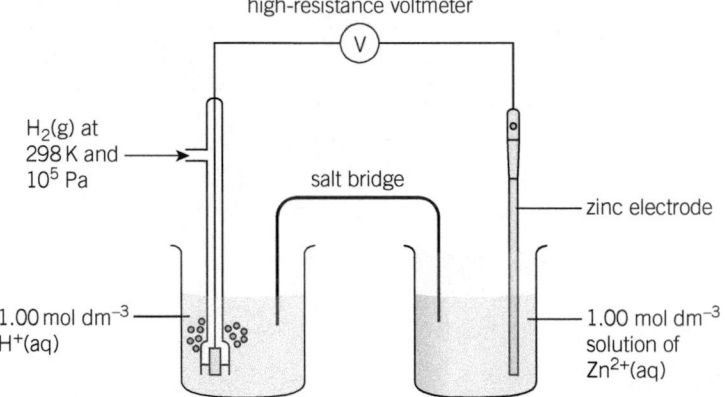

▲ **Figure 3** *An example of a standard electrochemical cell*

3 Check that the reading on the voltmeter is positive. This shows that the half-cell connected to the positive terminal of the voltmeter is the positive electrode.

4 If the reading is negative, change the connections round on the voltmeter to give a positive reading.

5 Record the voltmeter reading.

1 In the equation below, what is the change in oxidation state of S?

$$S_2O_3^{2-} + I_2 \rightarrow S_4O_6^{2-} + 2I^-$$

A +4 to +6   C +2 to +4

B +6 to +12   D 2– to 2– (no change in oxidation state)   (*1 mark*)

2 What is the correct full equation for the two half-equations given below?

$$V^{3+} + e^- \rightarrow V^{2+} \qquad E^\ominus = -0.26\,V$$
$$Zn^{2+} + 2e^- \rightarrow Zn \qquad E^\ominus = -0.76\,V$$

A $V^{2+} + Zn^{2+} \rightarrow Zn + V^{3+}$ C $V^{3+} + Zn^{2+} \rightarrow Zn + V^{2+}$

B $2V^{2+} + Zn \rightarrow Zn^{2+} + 2V^{3+}$ D $2V^{3+} + Zn \rightarrow Zn^{2+} + 2V^{2+}$ (*1 mark*)

3 What is E_{cell} for the vanadium/zinc cell given in question 2?

A 0.24 V   C 0.50 V

B 0.26 V   D 1.02 V   (*1 mark*)

4 What occurs at the more negative electrode of an electrochemical cell?   (*1 mark*)

A Oxidation, because electrons are gained

B Oxidation, because electrons are lost

C Reduction, because electrons are gained

D Reduction, because electrons are lost

5 Which of the following are correct explanations of different methods of preventing rusting?   (*1 mark*)

1 Oiling the item reduces contact with water

2 Painting the item reduces contact with atmospheric oxygen

3 Sacrificial protection involves more reactive metals corroding in preference to iron

A 1 only   C 3 only

B 1 and 2 only   D 1, 2, and 3

6 A student investigated the redox activity of vanadium compounds by converting VO_2^+ to VO^{2+} using zinc.

a Give the oxidation states of VO_2^+ and VO^{2+}.   (*2 marks*)

b Explain whether the vanadium was being oxidised or reduced.   (*1 mark*)

c Balance the half-equation for the conversion:
$$VO_2^+ + ...H^+ + ...e^- \rightleftharpoons VO^{2+} + ...H_2O$$ (*2 marks*)

d $E^\ominus$ for the half-reaction in part **c** is +1.00 V. $E^\ominus$ for Zn/Zn^{2+} is –0.76 V. Calculate E_{cell} for the reaction that the student investigated.   (*1 mark*)

e Write the balanced ionic equation for the redox reaction that the student carried out.   (*1 mark*)

f Vanadium can be further reduced to V^{3+} and V^{2+}. The half-cell values are $E^\ominus(VO^{2+}/V^{3+}) = +0.34\,V$ and $E^\ominus(V^{3+}/V^{2+}) = -0.26\,V$. Explain why zinc is able to reduce the vanadium species as far as V^{2+}.   (*1 mark*)

g $E^\ominus$ for Pb/Pb^{2+} is –0.13 V. What will be the final oxidation state of vanadium when VO_2^+ reacts with Pb? Give a reason.   (*2 marks*)

7 Give full, ionic, or half-equations for the following processes that occur during rusting:

a the oxidation of iron metal to Fe^{2+} ions   (*1 mark*)

b the reduction of oxygen in the presence of water   (*1 mark*)

c the formation of iron(II) hydroxide   (*1 mark*)

d the sacrificial oxidation of zinc metal   (*1 mark*)

10.5 Rates of reactions

Specification reference: CI(a), CI(c)

Synoptic link

You will learn more about using kinetics to propose the mechanism of a reaction in Topic 10.7, Finding the order of reaction with experiments.

Revision tip

The unit $mol\,dm^{-3}\,s^{-1}$ measures the change in concentration $(mol\,dm^{-3})$ per second (s^{-1}).

Synoptic link

You first learned about methods of measuring rates of reaction in Topic 10.1, Factors affecting reaction rates.

Synoptic link

You will learn about using concentrations to determine rate of reaction, including orders of reaction, in Topic 10.6, Rate equation of a reaction.

Key term

Quench: To effectively stop a reaction occurring by slowing down its rate suddenly.

Reaction kinetics

The study of reaction kinetics is an important area of chemistry. Understanding the rates of reactions allows chemists to identify factors which affect the rate, and gives insights into the mechanism of reactions.

During a reaction, reactants are used up as products are made. Rate of reaction is calculated by measuring the decrease in concentration of a reactant, or the increase in concentration of a product over a certain time period.

$$rate\ of\ reaction = \frac{change\ in\ concentration\ of\ reactant\ or\ product}{time\ taken}$$

The units of rate of reaction are usually $\mathbf{mol\,dm^{-3}\,s^{-1}}$.

To measure the rate of reaction it is necessary to identify a characteristic that can be easily monitored. These include:

- volume of gas, using a gas syringe
- mass changes, using a balance
- colour changes, using a colorimeter
- pH changes, using a pH meter.

These can be measured directly.

It is also possible to quench a reaction by adding a large quantity of water, rapidly cooling the reaction mixture, or by neutralising any acid reactant or catalyst. Then the concentration of one of the reactants or products can be determined by titration.

Summary questions

1 How is rate of reaction calculated? *(1 mark)*

2 Suggest how to measure the rate of reaction of magnesium reacting with hydrochloric acid. *(3 marks)*

3 Explain why quenching is necessary when continuous methods of measuring rates are not appropriate. *(1 mark)*

10.6 Rate equation of a reaction

Specification reference: CI(a), CI(d)

What is a rate equation?

A rate equation is a mathematical expression which enables the rate of reaction (in $mol\,dm^{-3}\,s^{-1}$) to be calculated.

For a general reaction in which A and B are the reactants:

$$A + B \rightarrow products$$

the rate equation is **rate = $k[A]^m\,[B]^n$**

[A] and [B] are the concentrations of reactants A and B, and m and n are the powers to which [A] and [B] are raised. m and n are the orders of reaction with respect to reactant A and B respectively. The overall order of reaction is $(m + n)$.

Orders of reactions usually have values of 0, 1, or 2. They indicate the effect of changing the concentration of the reactant on the rate of reaction.

It is very important to note that you cannot predict the rate equation for a reaction from its balanced equation. The only way of determining a rate equation is from experimental data.

Examples of rate equations

Iodine reacts with propanone as shown below. An acid catalyst (H^+) is used:

$$CH_3COCH_3(aq) + I_2(aq) \xrightarrow{H^+} CH_3COCH_2I(aq) + H^+(aq) + I^-(aq)$$

Experimental data show that the rate equation for this reaction is **rate = $k[CH_3COCH_3][H^+]$**

You can see the following with regard to the rate equation:

- It is first order with respect to CH_3COCH_3 because $[CH_3COCH_3]$ is raised to the power 1.
- It is first order with respect to H^+ – a catalyst can appear in the rate equation.
- It is zero order with respect to I_2 – changing the concentration of I_2 does not affect the rate of reaction, so $[I_2]$ does not appear in the rate equation.
- It is second order overall because the orders of reaction of individual components total 2.

A second example is the reaction of 2-bromo-2-methylpropane with hydroxide ions:

$$(CH_3)_3CBr(aq) + OH^-(aq) \rightarrow (CH_3)_3COH(aq) + Br^-(aq)$$

Experimental data show that the rate equation is **rate = $k[(CH_3)_3CBr]$**

Note the following:

- The reaction is first order with respect to $(CH_3)_3CBr$.
- It is zero order with respect to OH^-, as $[OH^-]$ does not appear in the rate equation.
- The overall order of reaction is first order.

Calculations involving rate equations

It is possible to calculate k by rearranging a rate equation. The units of k must

Key term

Overall order of reaction: The sum of the orders of reaction for each individual reactant.

▼ **Table 1** *Orders of reaction*

Order with respect to reactant	Effect of doubling the concentration of a reactant on the rate of reaction
zero	No effect (rate $\propto [A]^0$)
first	Rate is doubled when concentration is doubled (rate $\propto [A]$)
second	Rate is quadrupled when concentration is doubled (rate $\propto [A]^2$)

Revision tip

$\propto$ means 'is proportional to'.

Revision tip

A number raised to a power of zero is equal to 1.

Revision tip

Unlike equilibrium expressions such as K_c, rate equations *cannot* be determined from a chemical equation.

Synoptic link

You will find out more about determining orders of reaction in Topic 10.7, Finding the order of reaction with experiments.

Revision tip

A substance which is zero order does not appear in or before the rate-determining step of the reaction (see Topic 10.7, Finding the order of reaction with experiments).

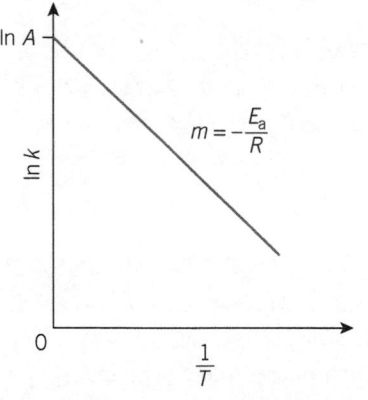

▲ **Figure 1** *Graph of* $\ln k$ *against* $\dfrac{1}{T}$.

Model answer: Calculating the value of k

The rate equation for the reaction of chloroethane with hydroxide ions is determined to be:

$$\text{rate} = k[C_2H_5Cl][OH^-]$$

Calculate k when $[C_2H_5Cl] = 0.002\,mol\,dm^{-3}$, $[OH^-] = 0.010\,mol\,dm^{-3}$, and rate $= 8.75 \times 10^{-2}\,mol\,dm^{-3}\,s^{-1}$. Give the units of k.

> Rearrange the rate equation with k as the subject.

$$k = \frac{\text{rate}}{[C_2H_5Cl][OH^-]}$$

> Substitute the values, then cancel units from the top and bottom.

$$k = \frac{8.75 \times 10^{-2}\,mol\,dm^{-3}\,s^{-1}}{0.002\,mol\,dm^{-3} \times 0.010\,mol\,dm^{-3}}$$

$$k = 4375\,\frac{mol\,dm^{-3}\,s^{-1}}{mol\,dm^{-3} \times mol\,dm^{-3}}$$

$$k = 4375\,mol^{-1}\,dm^3\,s^{-1}$$

also be calculated as they differ from equation to equation.

Rate constant k and temperature changes

$$\text{rate} = k[A]^m[B]^n$$

Increasing the temperature increases the rate of a chemical reaction. Since increasing the temperature does not affect $[A]$ or $[B]$, a rise in temperature must increase the value of the rate constant, k.

The Arrhenius equation

The Arrhenius equation is $k = Ae^{-\frac{E_a}{RT}}$, where

- k is the rate constant
- T is the temperature in kelvin
- R is the gas constant $8.314\,J\,K^{-1}\,mol^{-1}$
- E_a is the activation energy in joules per mole
- e is a mathematical constant.
- A is the frequency factor.

A, e, and R are constants, but E_a and T can vary. It is possible to use the Arrhenius equation to calculate E_a and A by taking the natural logarithm of both sides of the equation:

$$\ln k = \ln A - \frac{E_a}{RT}$$

This can be rearranged as $\ln k = -\dfrac{E_a}{R} \times \dfrac{1}{T} + \ln A$, which is in the form of the equation of a straight line, $y = mx + c$.

Therefore, plotting a graph of $\ln k$ on the y-axis against $\dfrac{1}{T}$ on the x-axis gives a straight line. The gradient of the line is $-\dfrac{E_a}{R}$ and the intercept is $\dfrac{1}{T}$.

Summary questions

1 Identify the orders of reaction with respect to each reactant, and the overall order of reaction, in the rate equation rate $= k\,[A]^2[B]$. (*3 marks*)

2 Explain how the order of reaction can be determined from a rate–concentration graph. (*3 marks*)

3 For a particular reaction, rate $= k[X]^2$. Calculate the value of k, including units, when $[X] = 0.01\,mol\,dm^{-3}$ and rate $= 4.2 \times 10^{-3}\,mol\,dm^{-3}\,s^{-1}$. (*3 marks*)

10.7 Finding the order of reaction with experiments

Specification reference: CI(b), CI(c), CI(e)

Experimental determination of reaction rate

It is not possible to deduce the rate equation from the balanced chemical equation for a reaction. Instead, it must be determined from a series of experiments.

The concentrations of the reactants [A], [B], etc., are varied one at a time, and the rate of reaction for those particular conditions is determined. The rate can be measured by measuring the change in a particular property, such as colour or pH.

The **initial rate** can be measured by plotting a graph of the change of this property over time, then drawing a tangent at time $t = 0$. The gradient of the tangent gives you the rate, as shown in the example below.

Revision tip

It is important to keep the temperature constant during the experiment, or k will change.

Synoptic link

Methods of measuring rates are discussed in Topic 10.1, Factors affecting reaction rates.

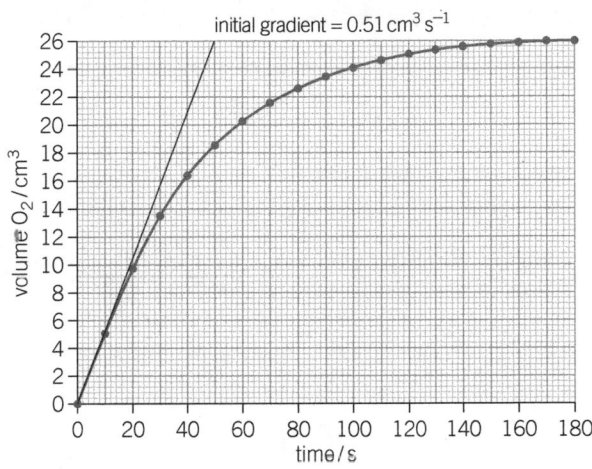

▲ **Figure 1** Drawing a tangent at time $t = 0$ to determine initial rate

The progress curve method can be used instead. A graph showing the change in concentration of a reactant is plotted over the course of the reaction. A series of tangents are taken at different concentrations. The gradient gives the rate of reaction at different times, as shown below.

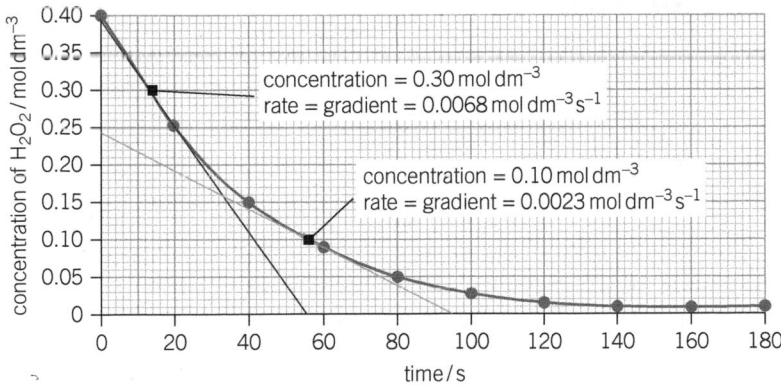

▲ **Figure 2** The progress curve method

Deducing the order of reaction for each reactant

Once a series of experiments are carried out, you will know how changing the concentration of a given reactant affects the rate of reaction. It is then possible to deduce the order of reaction:

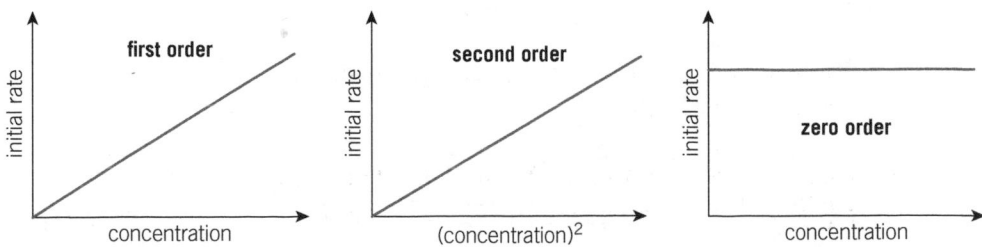

▲ **Figure 3** The effect of concentration on rate for first, second, and zero order reactants (note concentration² as the variable on the x axis in the second order graph)

55

Remember:

- If the rate is doubled when concentration is doubled (rate $\propto$ [A]), it is first order.
- If the rate is quadrupled when concentration is doubled (rate $\propto$ [A]^2), it is second order.
- If changing the concentration does not affect the rate, it is zero order.

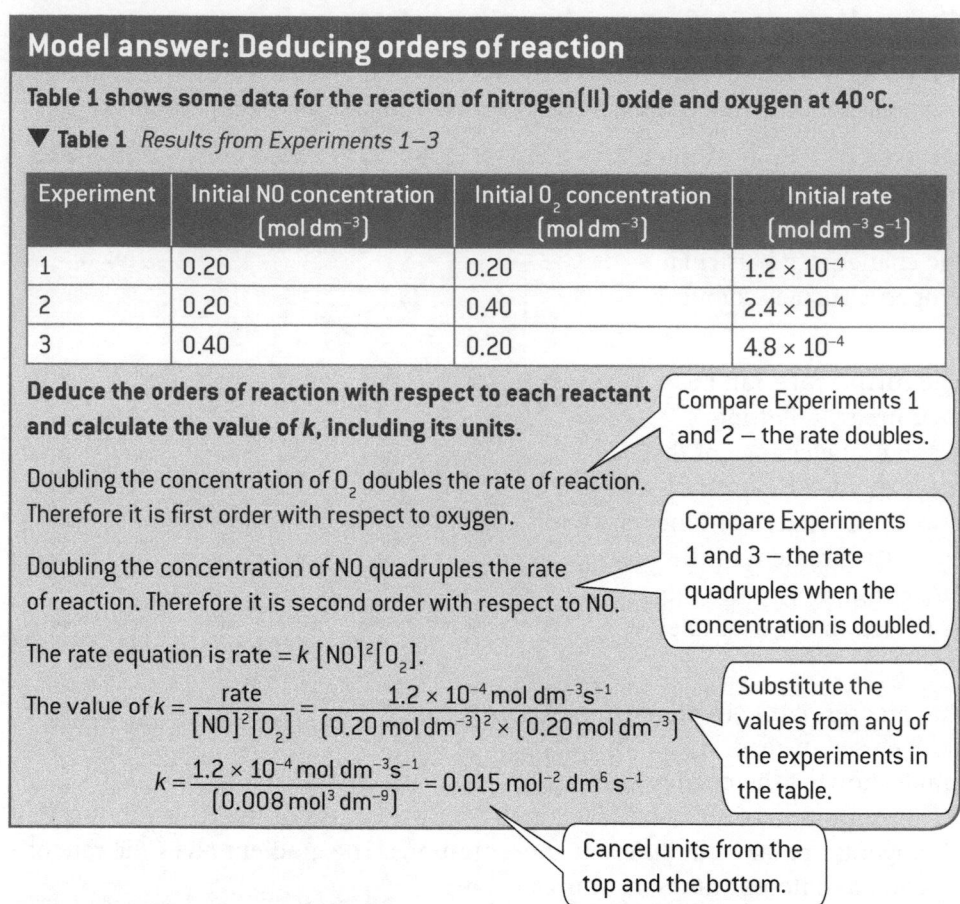

Model answer: Deducing orders of reaction

Table 1 shows some data for the reaction of nitrogen(II) oxide and oxygen at 40 °C.

▼ **Table 1** *Results from Experiments 1–3*

Experiment	Initial NO concentration (mol dm^{-3})	Initial O$_2$ concentration (mol dm^{-3})	Initial rate (mol dm^{-3} s^{-1})
1	0.20	0.20	1.2×10^{-4}
2	0.20	0.40	2.4×10^{-4}
3	0.40	0.20	4.8×10^{-4}

Deduce the orders of reaction with respect to each reactant and calculate the value of k, including its units.

Compare Experiments 1 and 2 – the rate doubles.

Doubling the concentration of O$_2$ doubles the rate of reaction. Therefore it is first order with respect to oxygen.

Compare Experiments 1 and 3 – the rate quadruples when the concentration is doubled.

Doubling the concentration of NO quadruples the rate of reaction. Therefore it is second order with respect to NO.

The rate equation is rate = k [NO]^2[O$_2$].

The value of $k = \dfrac{\text{rate}}{[\text{NO}]^2[\text{O}_2]} = \dfrac{1.2 \times 10^{-4}\,\text{mol dm}^{-3}\text{s}^{-1}}{(0.20\,\text{mol dm}^{-3})^2 \times (0.20\,\text{mol dm}^{-3})}$

Substitute the values from any of the experiments in the table.

$k = \dfrac{1.2 \times 10^{-4}\,\text{mol dm}^{-3}\text{s}^{-1}}{(0.008\,\text{mol}^3\,\text{dm}^{-9})} = 0.015\,\text{mol}^{-2}\,\text{dm}^6\,\text{s}^{-1}$

Cancel units from the top and the bottom.

The half-lives method

A rate–concentration graph shows how the concentration of a reactant changes over time. It is possible to use such a graph to determine the successive half-lives of the reactant, and this can be used to determine the order of reaction with respect to that reactant.

During the successive half-lives, the concentration decreases to half its original value, then one-quarter and then one-eighth. If the half-lives are a constant time, as in the graph below, the reaction is first order with respect to that reactant. Otherwise the reaction is second or zero order.

Key term

Half-life ($t_{1/2}$): The time taken for the concentration of a reactant to halve.

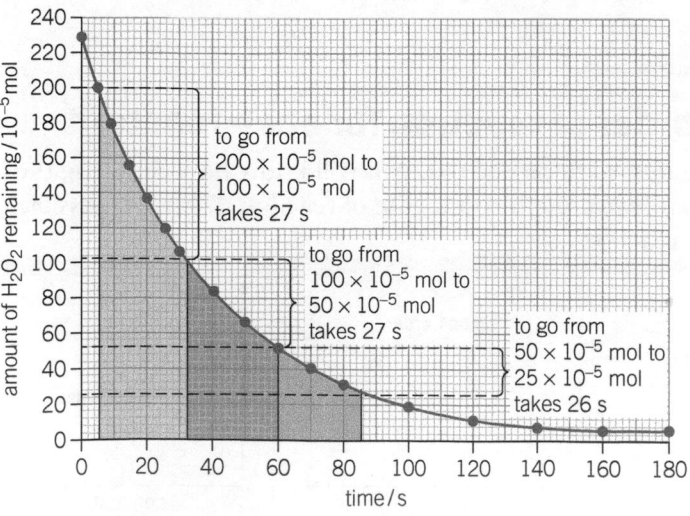

to go from 200×10^{-5} mol to 100×10^{-5} mol takes 27 s

to go from 100×10^{-5} mol to 50×10^{-5} mol takes 27 s

to go from 50×10^{-5} mol to 25×10^{-5} mol takes 26 s

▲ **Figure 4** *Calculating half-life from a graph*

Reaction mechanisms

By studying rate equations and orders, chemists can deduce a mechanism for a reaction. This describes the steps involved in the chemical reaction. The slowest step is known as the **rate-determining step**.

Rate-determining step

The rate equation tells us which particles are involved in the rate-determining step. For example:

$$(CH_3)_3CBr + OH^- \rightarrow (CH_3)_3COH + Br^-$$

rate = $k[(CH_3)_3CBr]$

The reaction is first order with respect to 2-bromo-2-methylpropane and zero order with respect to hydroxide ions. Therefore the rate-determining step involves only 2-bromo-2-methylpropane. It involves the heterolytic fission of the C–Br bond to form an intermediate:

$$(CH_3)_3CBr \rightarrow (CH_3)_3C^+ + Br^- \qquad \text{rate-determining step}$$

$$(CH_3)_3C^+ + OH^- \rightarrow (CH_3)_3COH \qquad \text{fast}$$

Another example is:

$$C_2H_5Br + OH^- \rightarrow C_2H_5OH + Br^-$$

rate = $k[C_2H_5Br][OH^-]$

In this case the rate-determining step involves both bromoethane and hydroxide ions. It occurs by nucleophilic substitution.

> **Key term**
>
> **Intermediate:** A chemical formed and then destroyed during the course of a reaction.

> **Key term**
>
> **Rate-determining step:** The slowest step of a reaction mechanism.

Summary questions

1 Sketch a graph showing successive half-lives for a first order reaction.

(3 marks)

2 Use the data below to determine the rate equation for the acid-catalysed reaction A + B ⟶ products. *(3 marks)*

[A] (mol dm^{-3})	[B] (mol dm^{-3})	[H$^+$] (mol dm^{-3})	Rate (mol dm^{-3} s^{-1})
0.01	0.01	0.01	4
0.02	0.01	0.01	8
0.02	0.02	0.02	16
0.02	0.02	0.04	32

3 For the reaction in question 2, a student proposes a two-step mechanism:
Slow step: A + H$^+$ ⟶ AH$^+$
Fast step: AH$^+$ + B ⟶ products
Explain whether or not this mechanism is consistent with the rate equation.

(1 mark)

10.8 Enzymes

Specification reference: PL(f), PL(g)

Key term

Substrate: A molecule that fits into the active site of an enzyme, and reacts with it.

Key term

Active site: A cleft in the enzyme surface where substrate molecules can bind and react.

What are enzymes?

Enzymes are proteins that act as catalysts in the body. An enzyme reacts with a substrate, and each enzyme has a high **specificity** for a particular substrate.

The active site of an enzyme is a region of the tertiary structure which is an exact fit for the structure of the substrate. All enzymes have an **active site**, where the tertiary structure of the enzyme exactly matches the structure of its substrate. The substrate binds to the active site, forming an enzyme–substrate complex. A reaction can then occur, after which the products leave the active site.

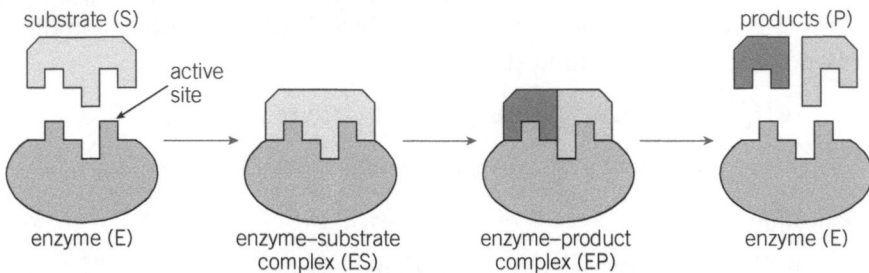

▲ **Figure 1** *A model of enzyme catalysis*

Sometimes a molecule that is a similar shape to the substrate binds strongly to the active site. This means that the usual substrate cannot enter, and the usual reaction does not occur. This is known as **competitive inhibition**.

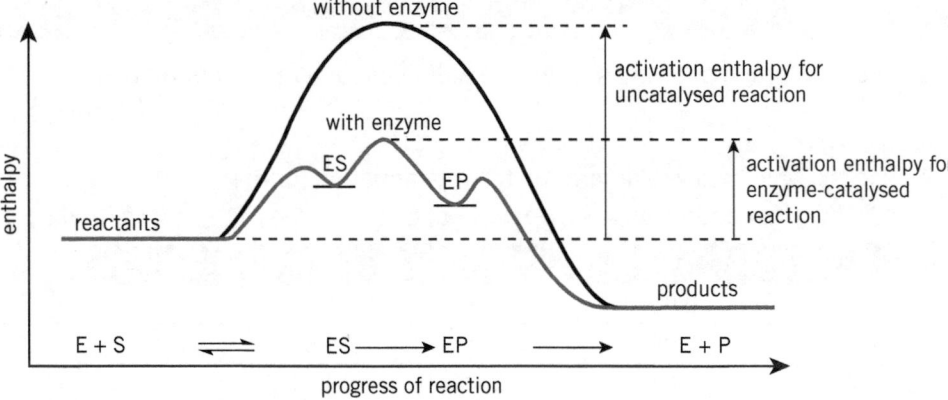

▲ **Figure 2** *How an enzyme provides a route of lower activation energy*

Kinetics of enzyme reactions

When the concentration of substrate is low, the rate equation is **rate = k[E][S]**. Under these conditions there are plenty of active sites for the substrate, so doubling the substrate concentration doubles the rate of reaction.

Synoptic link

Rate equations are covered in Topic 10.6, Rate equation of a reaction.

However when the substrate concentration is higher, at any given time all the active sites are occupied. The reaction is then zero order with respect to the substrate, as increasing the concentration of the substrate any further has no effect on the rate. Under these conditions, the rate equation is **rate = k[E]**.

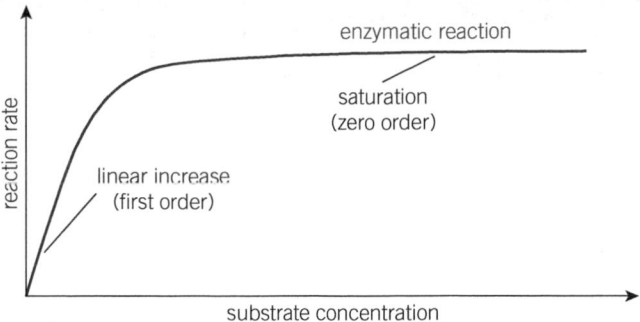

▲ **Figure 3** *How rate of reaction of an enzyme reaction varies with substrate concentration*

Changes to the shape of the active site

The shape of the active site depends on the tertiary structure of the protein that makes up the enzyme. The tertiary shape is determined by interactions between the amino acid side groups, such as hydrogen bonding and ionic bonds. Increasing the temperature can break hydrogen bonds and pH changes can disrupt ionic interactions; this leads to a change in the shape of the active site.

Consequently the enzyme is **denatured**. The substrate can no longer fit into the active site and the enzyme loses its activity.

Key term

Denatured: When the active site of the enzyme changes shape due to changes in pH, or increased temperature, the enzyme is denatured, and the substrate no longer fits.

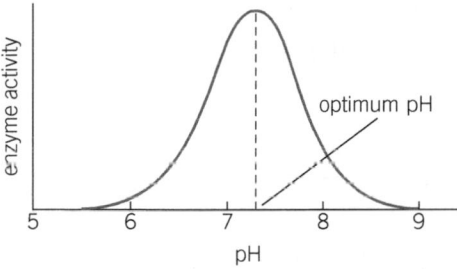

▲ **Figure 4** *Graph showing how the rate of a typical enzyme reaction varies with pH*

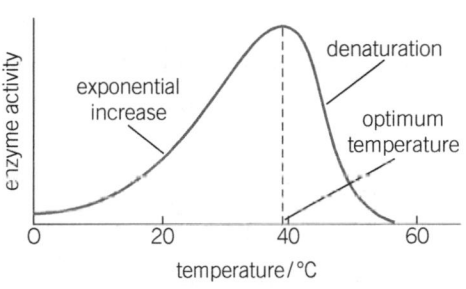

▲ **Figure 5** *Graph showing how the rate of an enzyme reaction varies with temperature*

Summary questions

1 What is: **a** a substrate **b** an inhibitor? (*2 marks*)

2 Why are enzymes denatured at high/low pH and at high temperatures? (*2 marks*)

3 Why does the rate equation for enzyme kinetics depend on the substrate concentration? (*1 mark*)

1 What are the units of rate of reaction?

   A there are no units

   C s^{-1}

   B it depends on the rate equation

   D $mol\,dm^{-3}\,s^{-1}$ *(1 mark)*

2 What is the overall order of reaction if rate = k [X]^2[Y]?

   A −1

   C 2

   B 1

   D 3 *(1 mark)*

3 What conclusions can be drawn from the rate equation rate = k [CH$_3$Br][OH$^-$]?

   1 The reaction is first order with respect to both reactants.

   2 The reaction is first order overall.

   3 The rate determining step is likely to involve CH$_3$Br and OH$^-$.

   A 1 and 2 only

   C 2 and 3 only

   B 1 and 3 only

   D 3 only *(1 mark)*

4 The statements below describe competitive inhibition of an enzyme by a molecule. Which is the most logical order of the statements? *(1 mark)*

   1 The molecule has a similar shape to the substrate.

   2 The molecule binds to the active site of the enzyme.

   3 The usual reaction cannot occur.

   4 The usual substrate cannot bind to the active site.

   A 4 − 2 − 3 − 1

   C 1 − 2 − 3 − 4

   B 1 − 2 − 4 − 3

   D 3 − 2 − 4 − 1

5 What is the rate equation for an enzyme reaction when the substrate concentration is high?

   A Rate = k

   C Rate = k [S]

   B Rate = k [E]

   D Rate = k [E] [S] *(1 mark)*

6 A particular reaction has the rate equation rate = k [M] [N]^2, where M and N are the reactants.

   a What will be the effect on the rate of doubling [M]? *(1 mark)*

   b What will be the effect on the rate of doubling both [M] and [N]? *(1 mark)*

   c Calculate the value of k, including its units, when rate = 0.025 mol dm^{-3} s^{-1} and [M] and [N] are both equal to 0.1 mol dm^{-3}. *(3 marks)*

   d What is the effect on k of increasing the temperature? *(1 mark)*

   e How will the half-lives of [M] and [N] differ in this reaction? (Assume that, in each case, all other reactants are in excess.) *(2 marks)*

7 Look at the table of data for the reaction: P + Q → S + T

[P] (mol dm^{-3})	[Q] (mol dm^{-3})	Rate (mol dm^{-3} s^{-1})
0.01	0.01	0.75 × 10^{-3}
0.02	0.01	1.50 × 10^{-3}
0.03	0.01	2.25 × 10^{-3}
0.03	0.02	4.50 × 10^{-3}

   a What is the order with respect to P and with respect to Q? *(2 marks)*

   b Explain how the initial rate could be determined from a graph of concentration [P] against time. *(1 mark)*

   c Comment on whether the following reaction mechanism, involving an intermediate R, is supported by the rate equation. *(1 mark)*

   Slow step: P → R + S

   Fast step: R + Q → T

11.4 The d-block

Specification reference: DM(g), DM(h), DM(i), DM(j), DM(k), DM(m), DM(n)

Oxidation states of transition metals

Transition metals often have variable oxidation states. They display a range of different oxidation states, which often are different in colour.

▼ **Table 1** *The common oxidation states of iron and of copper*

Oxidation state	Ion	Electron configuration	Colour of aqueous ion
Iron(II)	Fe^{2+}	$1s^2\,2s^2\,2p^6\,3s^2\,3p^6\,3d^6$	green
Iron(III)	Fe^{3+}	$1s^2\,2s^2\,2p^6\,3s^2\,3p^6\,3d^5$	orange/brown
Copper(I)	Cu^+	$1s^2\,2s^2\,2p^6\,3s^2\,3p^6\,3d^{10}$	N/A
Copper(II)	Cu^{2+}	$1s^2\,2s^2\,2p^6\,3s^2\,3p^6\,3d^9$	blue

Chromium and copper

In the ground state of an atom, electrons are arranged to give the lowest total energy. The energies of the 3d and 4s orbitals are very close together in period 4, and this leads to the electron configurations in Cr and Cu being different to what you might expect. In chromium, putting one electron in each 3d and 4s orbital gives a lower energy than having the usual two in 4s. In copper, putting two electrons in the 4s orbital would give a higher energy than filling up the 3d orbital.

- The electron configuration of Cr is **$1s^2\,2s^2\,2p^6\,3s^2\,3p^6\,3d^5\,4s^1$** (not $3d^4\,4s^2$).

- The electron configuration of Cu is **$1s^2\,2s^2\,2p^6\,3s^2\,3p^6\,3d^{10}\,4s^1$** (not $3d^9\,4s^2$).

Identifying transition metal ions

Fe^{2+}, Fe^{3+}, and Cu^{2+} ions can be readily identified through test tube reactions as they form precipitates when aqueous sodium hydroxide is added:

$Fe^{2+}(aq) + 2OH^-(aq) \rightarrow Fe(OH)_2(s)$ Iron(II) hydroxide is a green precipitate.

$Fe^{3+}(aq) + 3OH^-(aq) \rightarrow Fe(OH)_3(s)$ Iron(III) hydroxide is a brown precipitate.

$Cu^{2+}(aq) + 2OH^-(aq) \rightarrow Cu(OH)_2(s)$ Copper(II) hydroxide is a blue precipitate.

If ammonia solution is added, the same reactions occur, as ammonia is a source of OH^- ions. However, on addition of excess ammonia, the pale blue precipitate of $Cu(OH)_2$ then dissolves to give a deep blue solution of a copper/ammonia complex ion, $[Cu(NH_3)_4(H_2O)_2]^{2+}$. Iron(II) hydroxide and iron(III) hydroxide do not form complexes with ammonia.

Complex ions

A complex ion contains a transition metal ion surrounded by a number of ligands. Complex ions are written using square brackets, for example: $[Fe(H_2O)_6]^{2+}$, $[NiCl_4]^{2-}$, $[Ni(CN)_4]^{2-}$, or $[Ag(NH_3)_2]^+$.

Revision tip

The electron configuration of elemental iron is $1s^2\,2s^2\,2p^6\,3s^2\,3p^6\,3d^6\,4s^2$. The electron configuration of elemental copper is $1s^2\,2s^2\,2p^6\,3s^2\,3p^6\,3d^{10}\,4s^1$.

Revision tip

Remember when d-block metals lose electrons, they are removed from the 4s sub-shell before they are removed from the 3d sub-shell.

Synoptic link

You learned about electron configuration in Topic 2.3, Shells, sub-shells, and orbitals.

Revision tip

Remember to include state symbols for precipitation reactions.

Key term

Complex ion: A transition metal ion surrounded by a number of ligands.

Key term

Ligand: A negatively charged ion, or a neutral molecule with a lone pair of electrons, surrounding a transition metal ion.

Key term

Monodentate ligand: A ligand that attaches to a transition metal through one atom only.

Key term

Bidentate ligand: A ligand with two atoms with lone pairs or negative charges, which forms two bonds to a metal ion.

Key term

Polydentate ligand: A ligand which forms several bonds to a metal ion.

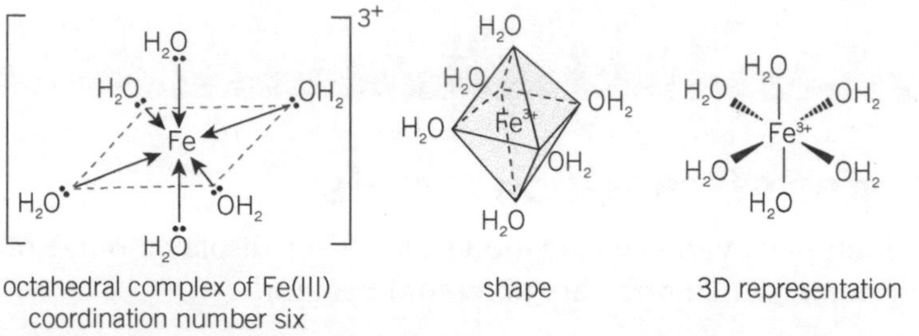

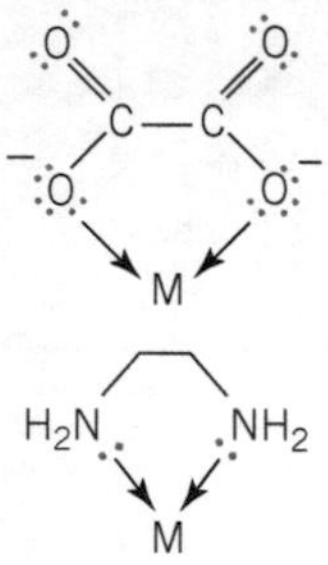

▲ **Figure 2** *Some bidentate ligands*

octahedral complex of Fe(III) coordination number six

shape

3D representation

▲ **Figure 1** *Some examples of complex ions*

Model answer: The charge of a complex ion

Work out the charge of the complex ion formed from **a** a copper(II) ion and six water molecules **b** a copper(II) ion and four chloride ions.

a   The copper ion has a charge of 2+.
The water molecules have no charge.

The total charge is $(2+) + (6 \times 0) = 2+$

Therefore the charge of the complex ion is 2+.

b   The copper ion has a charge of 2+.
The chloride ions have a charge of 1−.

The total charge is $(2+) + (4 \times 1-) = 2-$

Therefore the charge of the complex ion is 2−.

Monodentate, bidentate, and polydentate ligands

Ligands attach to transition metal ions through dative covalent bonding with the lone pairs on the ligands. Some ligands, such as H_2O and NH_3 can only bond through a single atom, and are called **monodentate ligands**. Some ligands, such as the ethanedioate ion and the 1,2-diaminoethane molecule, can form two dative covalent bonds to the metal ion, and are known as **bidentate ligands**. **Polydentate ligands** can form several dative covalent bonds as they contain several atoms with lone pairs or negative charges. $EDTA^{4-}$ is an example of a polydentate ligand.

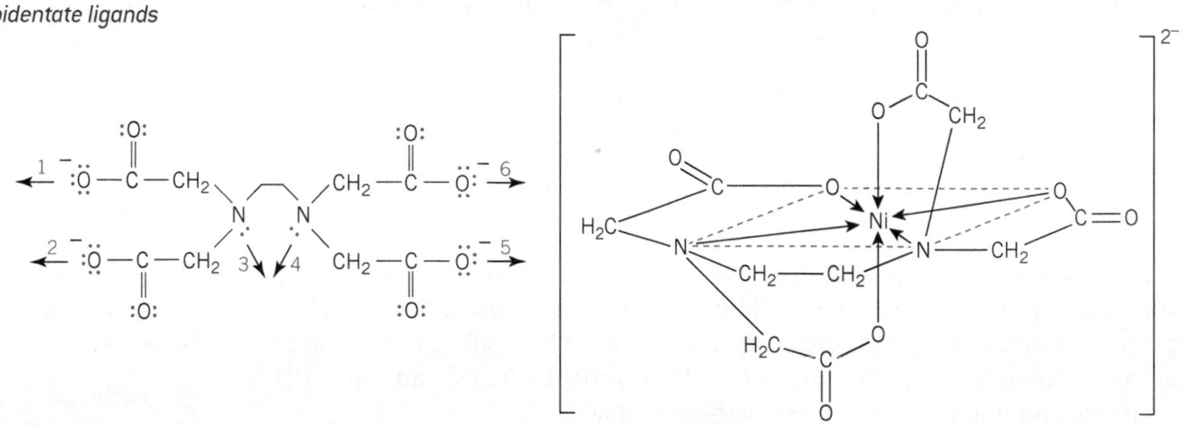

▲ **Figure 3** *A polydentate ligand ($EDTA^{4-}$)*

Coordination number

The number of bonds made between a metal ion and ligands is known as the **coordination number**. The most common coordination numbers are six and four. When a bidentate or polydentate ligand forms a complex, each metal–ligand bond is counted, so, for example, $[Ni(OOC-COO)_3]^{4-}$ has a coordination number of six.

▼ **Table 2** *The shapes of some complex ions*

Coordination number	Shape of complex	Example
6	octahedral	$[Fe(CN)_6]^{3-}$
4	tetrahedral	$[NiCl_4]^{2-}$
4	square planar	$[Ni(CN)_4]^{2-}$
2	linear	$[Ag(NH_3)_2]^+$

Colours of d-block ions

The presence of ligands causes the d sub-shell to split into higher and lower energy levels. When visible light is absorbed, electrons can be excited to a higher energy level. The frequency of light absorbed is proportional to the **gap** between the energy levels and is given by $\Delta E = h\nu$.

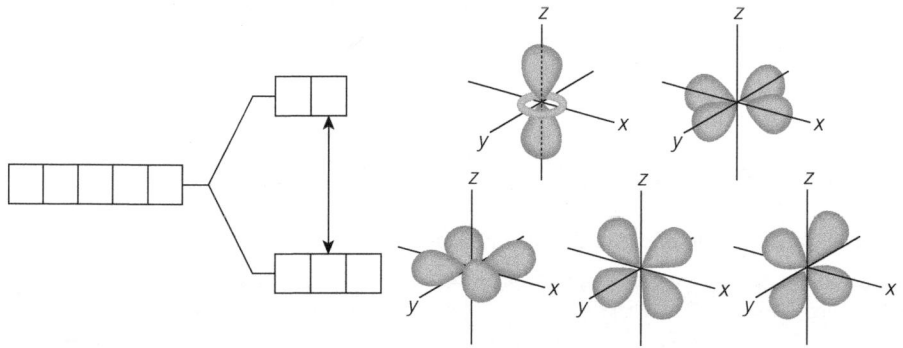

▲ **Figure 4** *Splitting of d-orbitals*

The complementary colour is transmitted, and is the colour we observe. For example, a Ti^{3+} ion absorbs yellow-green light, so it transmits the complementary colour, which is blue-violet. Therefore it appears a blue-violet colour.

Colorimetry

A colorimeter is used to find the concentration of a coloured solution. The absorbance of a solution is proportional to its concentration, so a more concentrated solution will absorb more strongly (see Experimental techniques).

Summary questions

1 Work out the charge of a complex ion formed from a cobalt(II) ion and four carbon monoxide molecules. *(1 mark)*

2 Explain how aqueous ammonia can be used to identify copper(II) ions. Give formulae for the species produced. *(4 marks)*

3 Explain how the presence of H_2O causes aqueous iron(III) ions to appear orange. *(4 marks)*

11.5 Nitrogen chemistry

Specification reference: Cl(j)

Go further: Nitrogen oxide

Nitrogen oxide, NO_2 is in equilibrium with a dimer, N_2O_4.

$$2NO_2 \rightleftharpoons N_2O_4 \quad \Delta H = -58.0\,kJ\,mol^{-1}$$

N_2O_4 is a pale yellow gas, whereas NO_2 is dark brown. Use ideas about changes in equilibrium position to predict what happens to the colour of a sample of NO_2 gas in a sealed container if it is cooled from 298 K to 273 K.

Nitrogen compounds

Nitrogen forms a range of compounds and ions. Many of these are important in soil chemistry: oxides such as N_2O, NO, and NO_2, oxoanions such as NO_3^-, NO_2^-, and ammonia, NH_3 and the ammonium ion NH_4^+.

Oxides of nitrogen

▼ **Table 1** *Formula and appearance of oxides of nitrogen*

Systematic name	Oxidation state of nitrogen	Molecular formula	Appearance
dinitrogen oxide	+1	N_2O	Colourless gas
nitrogen oxide	+2	NO	Colourless gas
nitrogen dioxide	+4	NO_2	Brown gas

Ammonia and ammonium ions

▼ **Table 2** *Bonding in ammonia and ammonium ions*

Name	Oxidation state of N	Formula	Dot-and-cross diagram	Shape of molecule or ion	Bond angle
Ammonia	−3	NH_3	H ∶N∶ H or more simply H—N—H H H	*Trigonal pyramid*	107°
Ammonium	−3	NH_4^+	H ∶N∶ H or more simply H—N⁺—H a dative covalent bond H H	*Tetrahedral*	$109\frac{1}{2}°$

Revision tip

Although you are only expected to **know** the dot-and-cross diagrams for ammonium and nitrogen (N_2), you should be able to *deduce* possible dot-and-cross diagrams for oxides and oxoanions of nitrogen, given suitable information about the species, such as the presence of multiple or dative bonds.

The lone pair of electrons on the N atom of the ammonia molecule allows it to act as a base. So it accepts H^+ ions from acids to become the ammonium ion.

$$NH_3 + H^+ \rightleftharpoons NH_4^+$$

Oxoanions of nitrogen (nitrate ions)

▼ **Table 3** *Name and formula of nitrate ions*

Name	Oxidation state of N	Formula
Nitrate(III)	+3	NO_2^-
Nitrate(V) [often just called nitrate]	+5	NO_3^-

Tests for ions containing nitrogen

All compounds containing nitrate(V) or ammonium ions are soluble. So it is not possible to use a test involving a precipitation reaction to positively identify these ions. Instead, tests involving the formation of ammonia gas are used.

Ammonium ions

Warm the substance being tested (solid or solution) with sodium hydroxide solution. Ammonia gas is formed. Ammonia gas causes moist red litmus paper, placed in the top of the test tube, to turn blue.

Synoptic link

Acid–base equilibria are covered in Topic 8.2, Strong and weak acids and pH.

Nitrate(V) ions

Sodium hydroxide is added to a solution containing a nitrate, followed by a spatula of powdered Devarda's alloy (a mixture of copper, aluminium, and zinc). Ammonia is formed (test using indicator paper as above). The aluminium metal reduces the nitrate(V) to ammonia.

Interconversion of nitrogen compounds

As shown in Tables 1–3, nitrogen can have oxidation states ranging from –3 to +5. Nitrogen compounds can take part in a variety of redox reactions, allowing ions and compounds containing nitrogen atoms to be interconverted. You can use ideas about oxidation states of nitrogen to write a half-equation for any of these interconversions. Half-equations may involve water molecules and H+ ions as well as electrons.

Revision tip

You should be able to use the dot-and-cross diagrams of the ammonia molecule and the ammonium ion to predict and describe the shapes of these particles.

Key term

An oxoanion is a negatively charged ion that contains one or more oxygen atoms bonded to a second element, such as sulfur or nitrogen.

Synoptic link

Tests for other common anions and cations are described in Topic 5.1, Ionic substances in solution.

 Worked example: Writing half-equations for the interconversion of nitrogen compounds

Balance this half-equation for the reduction of nitrate(V) ions to nitrogen(II) oxide:

$$NO_3^- + \_H^+ + \_e^- \rightarrow NO + \_H_2O$$

Step 1: Deduce the oxidation states of nitrogen in the reactant and product species: +5 in NO_3^-, +2 in NO.

Step 2: Calculate the number of electrons needed to cause this change in oxidation states: +5 to +2 requires the gain of $3e^-$.

So the equation becomes $NO_3^- + \_H^+ + 3e^- \rightarrow NO + \_H_2O$

Step 3: Balance the number of O atoms by adding in the appropriate number of H_2O molecules. There are 3O atoms in NO_3 and 1 in NO so there must be $2H_2O$:

$$NO_3^- + \_H^+ + 3e^- \rightarrow NO + 2H_2O$$

Step 4: Balance the number of H atoms in these water molecules by adding in the appropriate number of H^+ ions: There are 4H atoms in $2H_2O$, so $4H^+$ ions are needed.

$$NO_3^- + 4H^+ + 3e^- \rightarrow NO + 2H_2O$$

Step 5: Check that the charges balance on the RHS and LHS of the equation: right-hand side and left-hand side are both 0, so the equation is likely to be correct.

Many of these interconversions occur naturally in the soil and are important in maintaining the balance of nutrient ions such as nitrate(V) in the soil.

Summary questions

1   Give the systematic names for the following species:
    a N_2O b NO_2^- c NO_2 *(3 marks)*

2   A student suggests that a white solid might be ammonium nitrate. Outline how they could carry out chemical tests to confirm this.
                                                             (4 marks)

3   The formation of nitrate(V) ions is an important reaction that occurs in agricultural soils. Deduce a half-equation for the conversion of ammonium ions to nitrate(V) ions.   *(3 marks)*

Synoptic link

Combining two half-equations to show the overall equation for a redox reaction is explained in Topic 9.3, Redox reactions, cells, and electrode potentials.

Colorimetry

Colorimetry is used to determine the concentration of a coloured solution. Coloured solutions absorb specific wavelengths of light. The concentration is proportional to the amount of this light that is absorbed or transmitted.

- Make up a series of standard solutions of the test solution of known concentration.
- Select a filter with the complementary colour to the test solution.
- Zero the colorimeter with a tube of pure solvent.
- Measure the absorbance of the standard solutions and plot a calibration curve, showing concentration against absorbance.
- Measure the absorbance of the test solution and read off the concentration from the calibration curve.

A visible spectrophotometer can be used instead of a colorimeter. It works in a similar way except that it can give data for absorption or transmission for any given value in the visible spectrum. A colorimeter only gives information about a specific number of wavelengths, determined by the filter chosen.

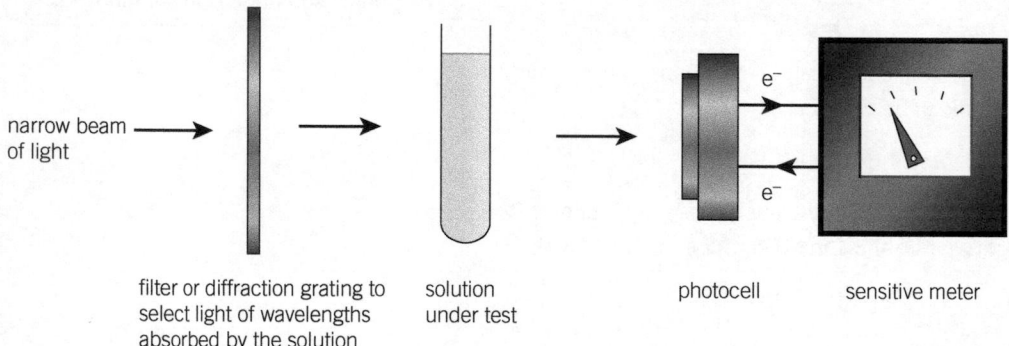

narrow beam of light

filter or diffraction grating to select light of wavelengths absorbed by the solution

solution under test

photocell

sensitive meter

▲ **Figure 1** *Colorimetry*

12.4 Arenes

Specification reference: CD (e)

Benzene

The simplest arene is benzene, C_6H_6.

Representing benzene

The structure and bonding in benzene can be represented in two ways:

- The first is based on the current model of benzene and uses a circle to show the delocalisation of electrons in the benzene ring.

- The second is based on the model of structure and bonding first proposed by Kekulé in the 19th century; this shows three separate double bonds in the molecule.

Delocalisation

The benzene ring contains delocalised electrons, represented by the circle in Figure 1 a.

The delocalisation can be described as follows:

- Six electrons are involved.

- These electrons come from the p orbitals of the C atoms in the benzene ring; one electron comes from each carbon.

- The six p orbitals overlap to form rings of electron density above and below the carbon atoms of the benzene ring.

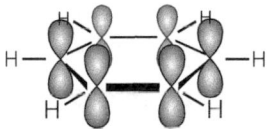 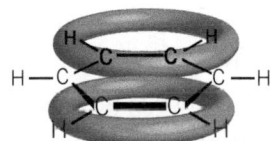

▲ **Figure 2** *The overlapping of p orbitals to form delocalised rings of electron density in a benzene ring*

The shape of the benzene molecule

- Benzene is a planar molecule.

- The carbon atoms are arranged in the form of a regular hexagon; all the C—C bond lengths are identical.

- All the bond angles in the molecule are 120°.

Relating shape to ideas about bonding

Both the Kekulé and the delocalised model of bonding would predict a planar hexagonal structure for benzene.

However, the Kekulé structure, with its alternating double and single bonds, would be an irregular hexagon with two different C–C bond lengths; with the C=C bond being shorter than the C–C bond.

The delocalised model, with electrons evenly spread over all the C atoms, correctly predicts the equal C–C bond lengths in the regular hexagonal structure.

Further evidence for the delocalised model

The Kekulé model suggests that the six electrons in p orbitals form three double bonds. Scientists used this idea, along with data for hydrogenation reactions of alkenes, to predict a value for the hydrogenation of benzene.

 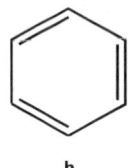

a   b

▲ **Figure 1** *The two representations of benzene* **a** *the delocalised structure* **b** *the Kekulé structure, showing three double bonds*

> ### Revision tip
> Benzene rings in questions may be shown using either the delocalised or the Kekulé representation. Unless you are told otherwise, you should use the delocalised representation when you are drawing benzene rings.

▲ **Figure 3** *The bond angles in benzene*

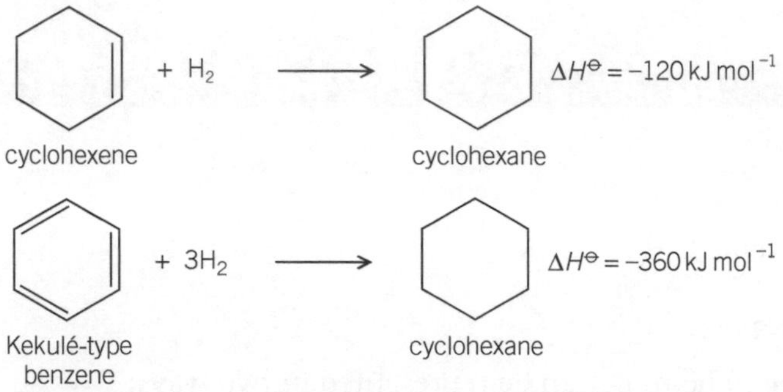

$$\text{cyclohexene} + H_2 \longrightarrow \text{cyclohexane} \qquad \Delta H^\ominus = -120\,\text{kJ mol}^{-1}$$

$$\text{Kekulé-type benzene} + 3H_2 \longrightarrow \text{cyclohexane} \qquad \Delta H^\ominus = -360\,\text{kJ mol}^{-1}$$

▲ **Figure 4** *The enthalpy changes for the hydrogenation of cyclohexene and the prediction for benzene using the Kekulé model*

The experimental value for benzene is $-208\,\text{kJ mol}^{-1}$, which is significantly less negative than the prediction using the Kekulé model.

This can be explained using the delocalised model. Delocalisation increases the strength of bonding in a molecule (and hence makes it more stable). More energy is needed to break the delocalised bonds and so the overall enthalpy change is less negative.

Evidence from the pattern of reactions for benzene

Benzene takes part in a range of substitution reactions and undergoes addition only under extreme conditions. This can be explained using the delocalised model; substitution reactions result in a product that also has a delocalised system and so the stability of this system is preserved.

The Kekulé model would be expected to undergo mainly addition reactions is it contains separate double bonds.

Aromatic compounds

Many derivatives of benzene can be synthesised, and the word **aromatic** is used to describe these compounds.

> ### ➕ Go further
>
> 'Aromatic' is strictly used to describe molecules that contain a cyclic structure with electrons that are delocalised in the same way as benzene.
>
> This delocalisation will occur if there are alternating double and single bonds and if there are $4n + 2$ electrons ($n = 1, 2$ etc.).
>
> Decide whether the structures in Figure 5 are aromatic or not:
>
>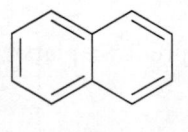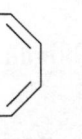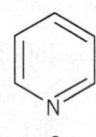
>
> a   b   c
>
> ▲ **Figure 5**

Naming arenes and aromatic compounds

Simple arenes and aromatic compounds

Substituted benzene rings can be named using the same ideas as cycloalkanes. The numbering of the carbon atoms starts at the group which comes first in alphabetical order.

Key terms

Arene: A hydrocarbon molecule that contains a benzene ring.

Aromatic compound: A molecule that contains a benzene ring.

Delocalised electrons: Electrons that are shared between more than two atoms.

Synoptic link

You also encountered the idea of delocalised electrons when studying the structure of metals in Topic 5.2, Bonding, structure, and properties.

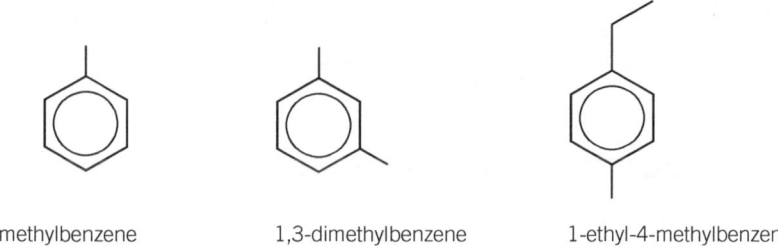

methylbenzene   1,3-dimethylbenzene   1-ethyl-4-methylbenzene

▲ **Figure 6** *Names of some of simple arenes*

Benzene rings containing halogens or nitro ($-NO_2$) groups occur commonly in reaction sequences.

More difficult names

Some names of aromatic compounds do not follow the simple naming pattern and are named in a less predictable way:

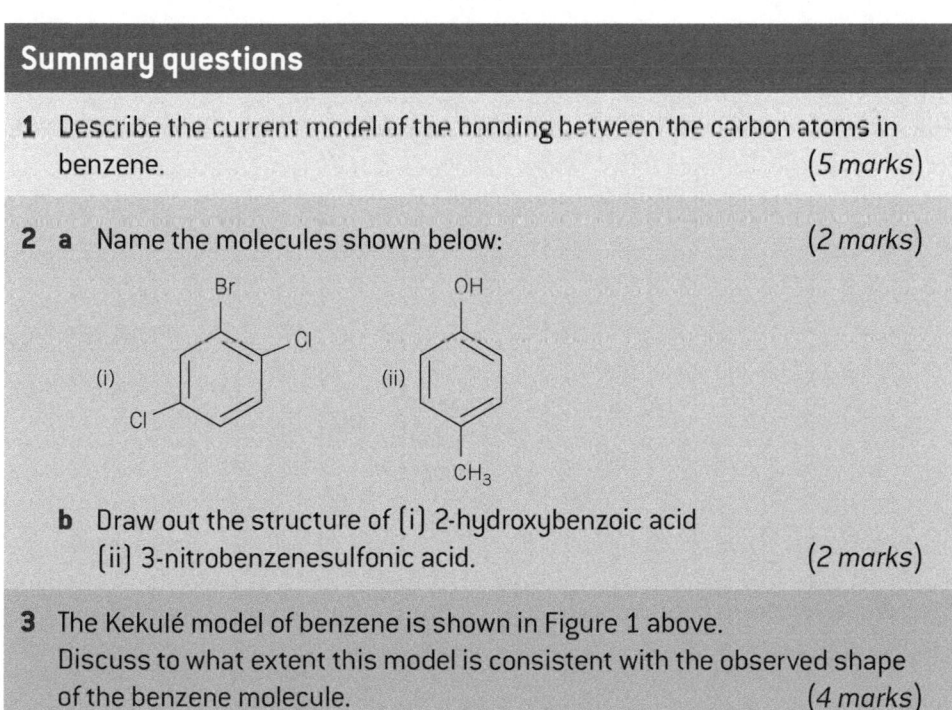

phenol   phenylamine   benzoic acid   benzene sulfonic acid

▲ **Figure 8** *The names of some other important aromatic compounds*

The phenyl group

Sometimes it makes more sense to think of the benzene ring as a side group in a structure. In these cases, the word 'phenyl' is used to describe the benzene ring.

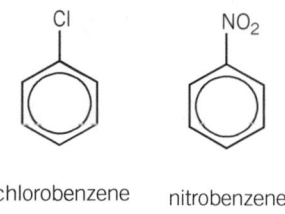

chlorobenzene   nitrobenzene

▲ **Figure 7** *Names of some of simple aromatic compounds*

Revision tip

You should learn the names of these common aromatic compounds, and be prepared to use other unpredictable names when they are introduced in a question.

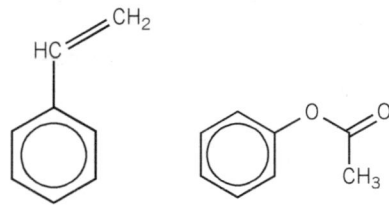

phenylethene   phenyl ethanoate

▲ **Figure 9** *Phenyl groups used in naming an alkene and an ester*

Synoptic link

If phenol is used to form esters, then the resulting ester will contain a phenyl group. The formation of esters is covered in Topic 13.4, Carboxylic acids and phenols.

Summary questions

1 Describe the current model of the bonding between the carbon atoms in benzene. *(5 marks)*

2 a Name the molecules shown below: *(2 marks)*

(i)   (ii)

   b Draw out the structure of (i) 2-hydroxybenzoic acid (ii) 3-nitrobenzenesulfonic acid. *(2 marks)*

3 The Kekulé model of benzene is shown in Figure 1 above. Discuss to what extent this model is consistent with the observed shape of the benzene molecule. *(4 marks)*

12.5 Reactions of arenes

Specification reference: CD (g)

Synoptic link

The definition of an electrophile is given in Topic 12.2, Alkenes.

Substitution reactions are described in Topic 13.2, Haloalkanes.

Revision tip

You should be able to compare the reactions of the delocalised benzene structure with those predicted by the **Kekulé** structure; this contains double bonds so addition reactions would be most likely. However it also has areas of high electron density so would also be attacked by electrophiles.

Synoptic link

The naming system for derivatives of benzene is explained in Topic 12.4, Arenes.

Common misconception: The molecular formula of the products of substitution reactions

If you are required to write the molecular formula of the products of substitution reactions, it is important to remember that the benzene ring has lost a H atom during the reaction – so the benzene ring shown in the structural formula now has formula C_6H_5, not C_6H_6.

Electrophilic substitution reactions

The benzene ring in arenes and other aromatic compounds generally takes part in electrophilic substitution reactions, although the reactions are often quite slow.

Explaining the reactions of benzene

The reactions of the benzene ring can be explained using ideas about the delocalised structure of benzene:

● Benzene contains an area of high electron density (the ring of delocalised electrons), so it is likely to be attacked by electrophiles.

● Substitution reactions (replacing one of the H atoms from the benzene ring with a different atom or group of atoms) does not affect the delocalised system of electrons, so the product is relatively stable.

● Benzene does not usually take part in addition reactions, because addition would destroy the delocalised system of electrons.

Because the rate of reaction can be very slow, specific catalysts often need to be present, as well as certain key conditions. These are necessary in order to generate a species that is reactive enough to act as an electrophile.

Halogenation reactions

Benzene reacts with bromine or chlorine in the presence of certain catalysts to form halogenated derivatives of benzene.

▼ Table 1 *Halogenation reactions of benzene*

Reaction	Reagent	Catalyst	Conditions	Electrophile	Product
bromination	bromine	iron (or iron(III) bromide)	reflux	Br^+	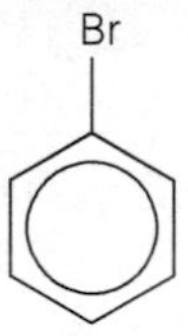 Br ⬡ bromobenzene
chlorination	chlorine	aluminium chloride	room temperature, anhydrous conditions	Cl^+	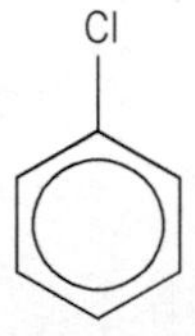 Cl ⬡ chlorobenzene

Equations for these reactions

The overall equation for the bromination of benzene is:

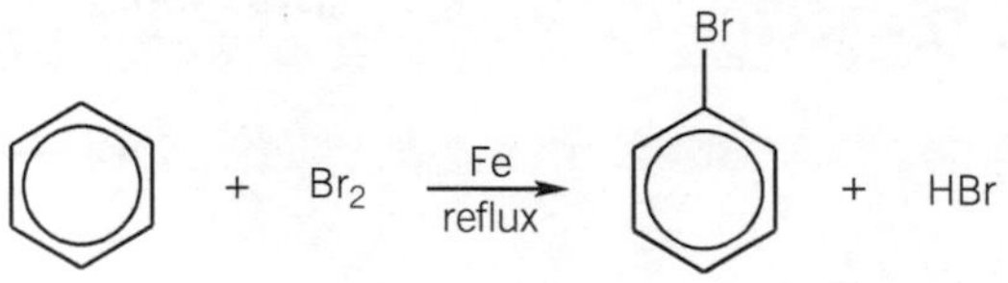

▲ Figure 1 *Equation for the bromination of benzene*

It can sometimes be helpful to show the process that occurs with the actual electrophile (Br^+):

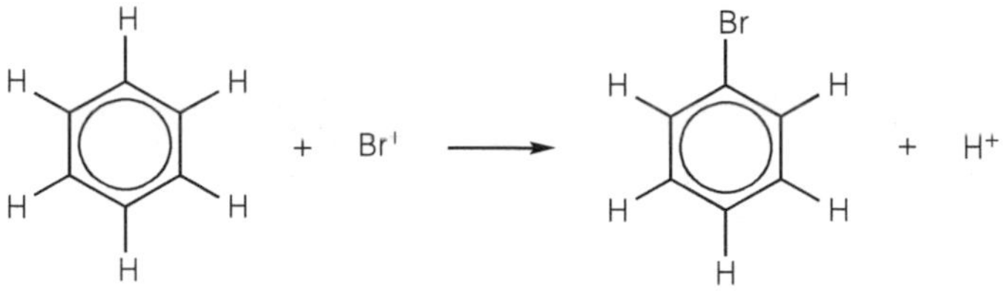

▲ **Figure 2** *The reaction that occurs with the Br^+ electrophile*

The equations for chlorination are very similar to those for bromination.

Nitration and sulfonation reactions

Nitration and sulfonation reactions are important in modifying the properties of dye molecules based on benzene rings.

▼ **Table 2** *Nitration and sulfonation reactions*

Reaction	Reagent	Catalyst	Conditions	Electrophile	Product
nitration	conc. nitric acid	conc. sulfuric acid	below 55 °C (to avoid multiple nitrations)	NO_2^+	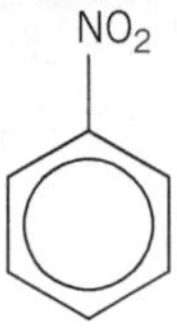 NO_2 nitrobenzene
sulfonation	conc. sulfuric acid	none required	reflux	SO_3	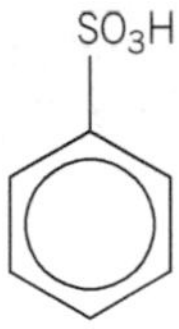 SO_3H benzene sulfonic acid

Equations for these reactions

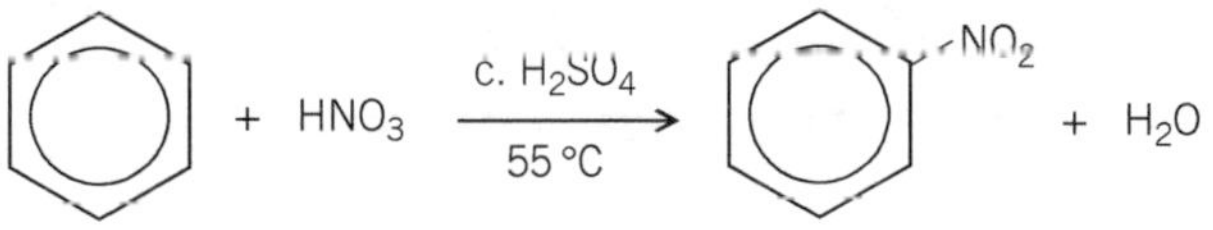

▲ **Figure 3** *The equation for nitration of benzene*

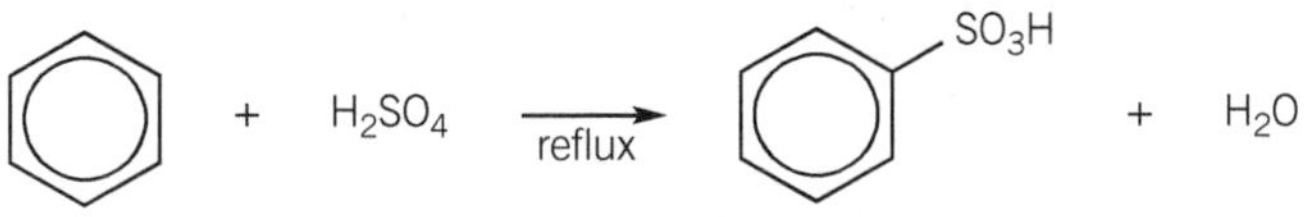

▲ **Figure 4** *The equation for sulfonation of benzene*

Mechanism for the nitration reaction

The NO_2^+ is generated by the reaction between the sulfuric acid catalyst and nitric acid.

This NO_2^+ ion acts as an electrophile, accepting a pair of electrons from the benzene ring to form a positively charged intermediate.

This intermediate is unstable. A pair of electrons from a C–H bond becomes part of the delocalised system and an H^+ ion is lost from the intermediate.

Finally, the H^+ ion causes the regeneration of the sulfuric acid catalyst.

The mechanism is shown in Figure 5.

➕ Go further – Halogen carriers

The catalysts used in halogenation are involved in the formation of the highly electrophilic Br^+ and Cl^+ ions.

For example, $AlCl_3$ accepts a Cl^- ion from a chlorine molecule, leaving behind a Cl^+ ion.

The H^+ released from the substitution reaction then regenerates the $AlCl_3$ catalyst. Catalysts that act in this way are known as **halogen carriers**.

The $FeBr_3$, which is formed when an Fe catalyst reacts with bromine, acts in a similar way.

$AlCl_3$ acts as a halogen carrier catalyst in the presence of Cl_2.

Write equations for: **a** the formation of Cl^+ **b** the regeneration of the $AlCl_3$ catalyst.

Synoptic link

The structure and properties of dye molecules are described in Topic 12.6, Azo dyes.

Step 1: $HNO_3 + H_2SO_4 \rightarrow NO_2^+ + HSO_4^- + H_2O$

Step 2:

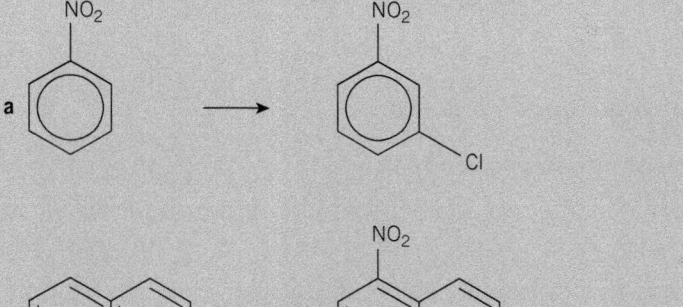

intermediate

Step 3: $H^+ + HSO_4^- \rightarrow + H_2SO_4$

▲ **Figure 5** *The mechanism of nitration*

Friedel–Crafts reactions

Alkyl groups (such as ethyl groups $-CH_2CH_3$) or acyl groups (such as ethanoyl groups $-COCH_3$) can be attached to a benzene ring using Friedel–Crafts reactions (named after the two 19th century chemists who developed these processes).

▼ **Table 3** *Products of Friedel–Crafts reactions*

Reaction	Reagent	Catalyst	Conditions	Electrophile	Product
alkylation	haloalkane e.g. CH_3CH_2Cl	aluminium chloride, $AlCl_3$	anhydrous conditions, heat under reflux	$CH_3CH_2^+$	CH_2CH_3 ethylbenzene
acylation	acyl chloride e.g. CH_3COCl	aluminium chloride, $AlCl_3$	reflux	SO_3	SO_3H benzene sulfonic acid

Summary questions

1 Explain why benzene takes part in substitution reactions, rather than addition reactions. *(3 marks)*

2 Give the reagents and conditions necessary for these reactions: *(4 marks)*

a

b

3 Benzene can react with propanoyl chloride (CH_3CH_2COCl) in a Friedel–Crafts reaction:
   a State what other substance needs to be present and explain the role of this substance. *(3 marks)*
   b Write an equation, using structural formulae, for this reaction. *(3 marks)*

12.6 Azo dyes

Specification reference: CD (b), CD (h)

Azo dyes

A dye is a coloured molecule that can attach to fabrics and other solid substance. Many dye molecules are described as azo dyes because they contain an azo group.

Formation of azo dyes

Azo dyes are synthesised from aromatic amines in two stages:

- A diazonium compound is formed as an intermediate.

- The diazonium compound is then coupled to a second aromatic molecule.

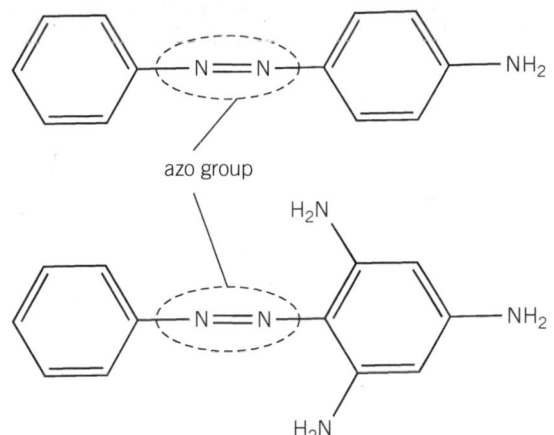

▲ **Figure 1** *The structure of some azo dyes, showing the azo functional group*

Key term

Azo group: An organic functional group with the structure $R_1-N=N-R_2$.

Key term

Diazonium compound (or ion): A compound or ion that contains a diazo functional group ($R-N^+\equiv N$).

Forming a diazonium compound

Aromatic amines, such as phenylamine, react with nitric(III) acid, HNO_2, in the presence of hydrochloric acid.

The nitric(III) acid is usually formed in the reaction mixture by adding a solution of sodium nitrate(III) to hydrochloric acid.

This is added to a solution of phenylamine in hydrochloric acid.

The temperature must be kept below 5 °C to slow down the rate at which the diazonium compound decomposes.

$$NH_2 + HNO_2 + H^+ \longrightarrow -N^+\equiv N + 2H_2O$$

▲ **Figure 2** *The equation for the formation of a diazonium ion*

Coupling reactions

The unstable diazonium compound must be used as soon as possible in the second stage of the synthesis.

The diazonium ion reacts with aromatic compound, called a coupling agent. This coupling agent is usually a phenol, or another aromatic amine:

- A solution of the coupling agent is made up.

- An ice-cold solution of the diazonium compound is added to it.

- A coloured suspension of the azo dye forms immediately.

This reaction is an electrophilic substitution reaction because the diazonium ion acts as an electrophile and replaces one of the H atoms on the benzene ring of the coupling agent.

The presence of the phenol or amine group increases the reactivity of the coupling agent by increasing the electron density on the benzene ring. When phenols are used as coupling agents they are usually dissolved in sodium hydroxide to form a negative phenoxide ion which increases the electron density even further.

Revision tip

You may need to predict a possible structure for the azo dye formed from a given combination of aromatic amine and coupling agent, or to work backwards to deduce the structures of the starting materials needed to form a particular azo dye.

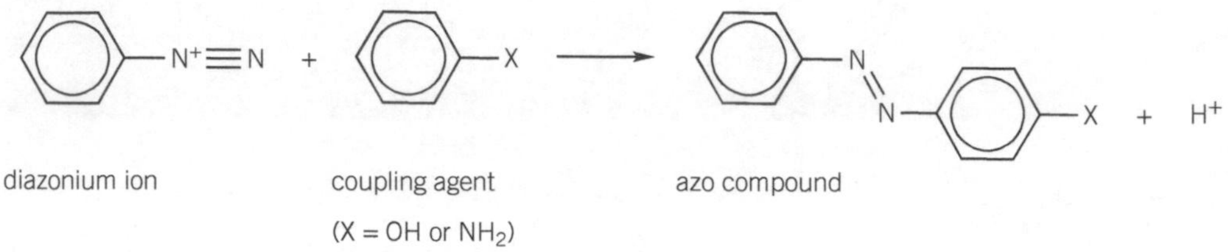

▲ Figure 3 *The equation for a coupling reaction, forming an azo dye*

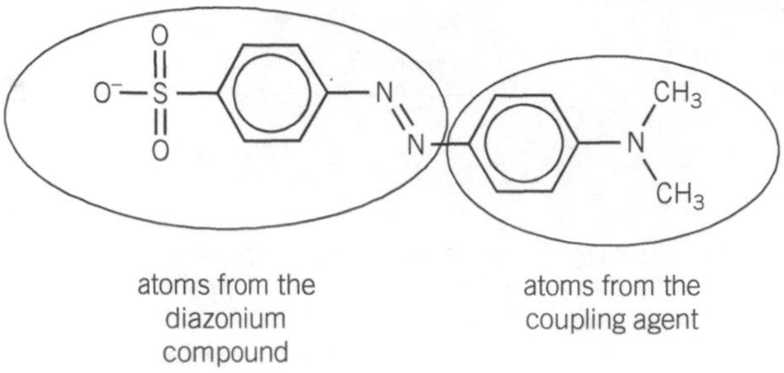

▲ Figure 4 *The structure of methyl orange, showing the parts of the molecule that originated in the diazonium compound and the coupling agent*

Remember that one of the nitrogen atoms in the azo group comes from the aromatic amine; the other one comes from the nitrogen atom from the nitric(III) acid. The coupling agent will retain its OH or amine group, even after the coupling reaction.

Chromophores in azo dyes

Azo dyes usually have intense orange or yellow colours because the delocalised system of the chromophore is extensive enough to absorb visible light in the blue part of the spectrum.

Modifying dye molecules

Chemists modify the structure of dyes by substituting various functional groups into the benzene rings of the azo dye.

Modifying the chromophore

The chromophore can be modified by including groups of atoms, such as nitro groups ($-NO_2$), that extend the delocalised system. This will alter the colour of the dye.

Increasing solubility

Many dyes need to be water soluble in order to be used in the dyeing process. Sulfonic acid groups ($-SO_3H$) in the dye structure increase solubility because these groups can easily be converted into the ionic sulfonate group ($-SO_3^-$).

Bonding dye molecules to fibres

Groups such as sulfonic acid (SO_3H), hydroxyl (OH), or amine (NH_2) are also introduced into dye molecules to allow them to bond more effectively to fabric.

Synoptic link

The concept of a chromophore was introduced in Topic 6.9, Coloured organic molecules.

Synoptic link

Substituting nitro- and sulfonic acid groups into benzene rings was described in Topic 12.5, Reactions of arenes.

Synoptic link

The bonding of dyes to fabric is described in detail in Topic 5.7, Bonding dyes to fibres.

Summary questions

1 Describe how phenylamine, $C_6H_5NH_2$, can be converted into the diazonium compound phenyldiazonium chloride, $C_6H_5N_2^+Cl^-$. *(3 marks)*

2 The structure of the azo dye acid black 1 is shown:

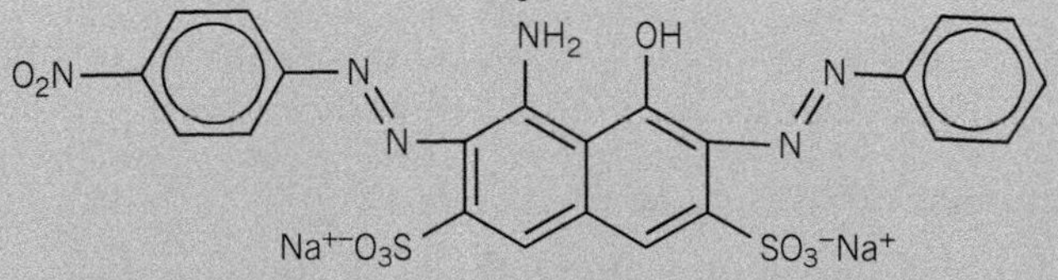

Suggest possible reasons for the inclusion of the following groups in the structure of this dye molecule:

a the $SO_3^-Na^+$ groups   c the NO_2 (O_2N) group
b the OH and NH_2 groups. *(3 marks)*

3 The azo dye shown below is synthesised from two organic starting materials:

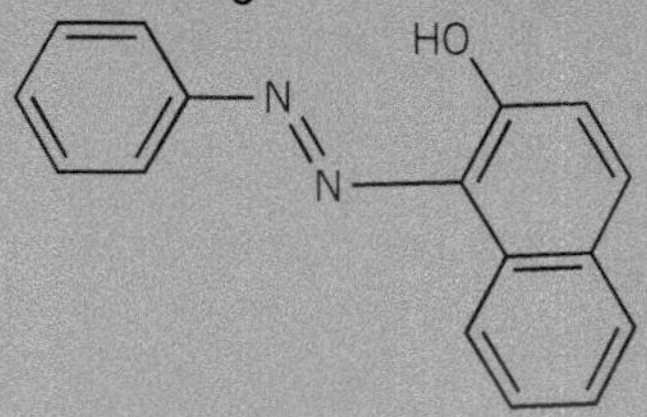

Deduce the structures of the two organic starting materials. *(2 marks)*

Chapter 12 Practice questions

1 The molecular formula of benzenesulfonic acid is:

   A $C_6H_5SO_3H$ **B** $C_6H_5SO_3$ **C** $C_6H_7SO_4$ **D** $C_6H_6SO_3$ (*1 mark*)

2 Nitro groups ($-NO_2$) are sometimes substituted into a benzene ring in a dye molecule. The most likely reason for this is:

   A To alter the wavelength at which the dye absorbs visible light.

   B To increase the solubility of the dye.

   C To improve the ability of the dye to bond to fibres.

   D To make the dye more stable.   (*1 mark*)

3 Which two substances would react in a coupling reaction to form the dye molecule shown (left)?

   A Benzene diazonium chloride and 1,3-dihydroxybenzene

   B Phenylamine and 1,3-dihydroxybenzene

   C Benzene and 1,3-dihydroxybenzene diazonium chloride

   D Phenylamine and 1,3-dihydroxybenzene   (*1 mark*)

4 Which of the following intermediates are involved in the nitration of benzene by a mixture of concentrated nitric and sulfuric acids

   1 $C_6H_6NO_2^+$ **2** NO_2^+ **3** $C_6H_5^+$

   A 1,2, and 3   **B** 1 and 2   **C** 2 and 3   **D** Only 3   (*1 mark*)

5 The reaction between benzene and 1-chloropropane to produce propyl benzene can be described as:

   1 Electrophilic substitution   **2** A Friedel-Crafts reaction   **3** Acylation

   A 1,2, and 3   **B** 1 and 2   **C** 2 and 3   **D** Only 3   (*1 mark*)

6 The structure and bonding in benzene can be represented by two different models, as shown (left).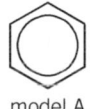

   a Describe how the bonding differs in these two models.   (*2 marks*)

   model A   model B

   b Discuss the extent to which experimental evidence is consistent with these two models.   (*6 marks*)

7 The following sequence of reactions shows a four-step synthesis of a dye molecule.

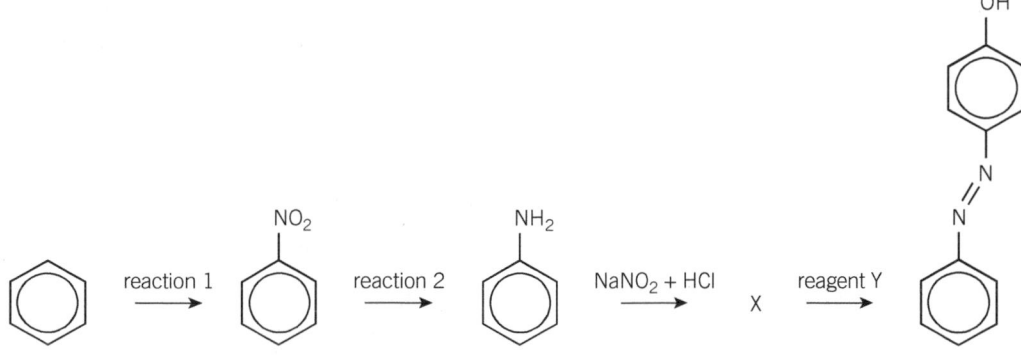

   a State the reagents that would be required to carry out:

     i Reaction 1   **ii** Reaction 2   (*2 marks*)

   b Draw out the structure of:

     i Molecule X   **ii** Reagent Y   (*2 marks*)

   c Reaction 1 and the reaction that forms X are both carried out at particular conditions of temperature. For **each** reaction, state the temperature that is required and explain why it is necessary.   (*4 marks*)

13.5 Carboxylic acids

Specification reference: PL(h), PL(p)

The structure of carboxylic acids

As you saw in Topic 13.4, Carboxylic acids and phenols, carboxylic acids contain the –COOH functional group, known as the carboxyl group. When counting carbons in the longest chain to name a carboxylic acid, the carbon in the carboxylic acid group is included. For example, CH_3CH_2COOH is propanoic acid. When two carboxyl groups are present the ending is **-dioic acid**. In this case the e in the name of the alkane is kept, e.g. $HOOCCH_2COOH$ is propanedioic acid.

Reactions of carboxylic acids

In aqueous solution, carboxylic acids exist in equilibrium with their dissociated ions:

$$RCOOH(aq) + H_2O(l) \rightleftharpoons RCOO^- (aq) + H_3O^+ (aq)$$

Carboxylic acids are weak acids. They are only partially dissociated in solution.

The reaction with metals

Carboxylic acids react with metals to form a salt and hydrogen. The general reaction is:

$$RCOOH + M \rightarrow RCOO^- M^+ + \frac{1}{2}H_2$$

For example, magnesium reacts with benzoic acid as follows:

$$2C_6H_5COOH(aq) + Mg(s) \rightarrow (C_6H_5COO)_2Mg (aq) + H_2(g)$$

The reaction with bases

Carboxylic acids react with bases to form a salt and water. Many useful derivatives can be made from the salts of carboxylic acids. The general reaction is:

$$RCOOH + MOH \rightarrow RCOO^- M^+ + H_2O$$

For example, potassium hydroxide reacts with propanoic acid as follows:

$$C_2H_5COOH (aq) + KOH (aq) \rightarrow C_2H_5COOK (aq) + H_2O (l)$$

The reaction with carbonates

Carboxylic acids react with carbonates to form a salt, carbon dioxide, and water. This is the basis of a test for carboxylic acids – they fizz when added to sodium carbonate. The reaction produces bubbles of carbon dioxide gas, which are readily seen and can be confirmed by testing the gas with limewater, which forms a cloudy precipitate.

The general equation for the reaction of carboxylic acids with carbonates is:

$$RCOOH + M_2CO_3 \rightarrow RCOO^- M^+ + CO_2 + H_2O$$

For example, sodium carbonate reacts with ethanoic acid as follows:

$$2CH_3COOH(aq) + Na_2CO_3(s) \rightarrow 2CH_3COONa(aq) + CO_2(g) + H_2O(l)$$

Synoptic link

You learned about weak acids in Topic 8.2, Strong and weak acids and pH.

Revision tip

Remember, a magnesium ion is Mg^{2+}, and carboxylate ions are $RCOO^-$.

Revision tip

The test for carbon dioxide involves the reaction of limewater, $Ca(OH)_2$, with CO_2:

$$Ca(OH)_2 (aq) + CO_2 (g) \rightarrow$$
$$CaCO_3 (s) + H_2O (l)$$

$CaCO_3$ forms a cloudy precipitate.

Revision tip

In this example, M is an ion with a single + charge, such as Na^+.

Summary questions

1 Name the following carboxylic acids:

   a $CH_3CH_2CH_2COOH$ **b** $HOOCCH_2CH_2COOH$ **c** C_6H_5COOH *(3 marks)*

2 Write an equation for the reaction of ethanoic acid with:

   a potassium   **b** potassium hydroxide   **c** potassium carbonate.   *(3 marks)*

3 A student reacts propanoic acid with two substances, X and Y. The reaction with X produces a colourless gas that burns with a squeaky pop, and the reaction with Y produces a colourless gas that turns limewater cloudy. What type of substances are X and Y?   *(2 marks)*

13.6 Amines

Specification reference: PL(j), PL(k), PL(l), PL(n)

Formulae and nomenclature of amines

Amines are organic compounds, derived from ammonia, NH_3. In amines, one or more of the hydrogens in ammonia are substituted by alkyl groups. The formulae of amines are therefore RNH_2, R_2NH, or R_3N.

Primary, secondary, and tertiary amines

In primary amines, there is one R group. In secondary amines, there are two R groups, and in tertiary amines, there are three R groups.

Naming primary amines

Primary amines are named by taking the name of the R group and adding the suffix 'amine'. For example, CH_3NH_2 is methylamine, $CH_3CH_2NH_2$ is ethylamine, $CH_3CH_2CH_2NH_2$ is propylamine, and $C_6H_5NH_2$ is phenylamine.

Where the amino group is attached to the middle of a chain, numbers are required to indicate its position. For example, $CH_3CH(NH_2)CH_3$ is 2-propylamine.

Diamines

If there are two $-NH_2$ groups, the prefix 'diamino' is used. For example, 1,6-diaminohexane is $H_2NCH_2CH_2CH_2CH_2CH_2CH_2NH_2$. It is one of the monomers for the production of nylon-6,6.

Amines as bases

The nitrogen atom on the NH_2 group has a lone pair of electrons. This is responsible for much of the chemistry of amines. Amines act as bases, as they can accept protons to form a dative covalent bond.

When amines dissolve in water, they form alkaline solutions:

$$RNH_2 + H_2O \rightarrow RNH_3^+(aq) + OH^-(aq)$$

Amines can accept a hydrogen ion from an acid. The general reaction is:

$$RNH_2 + H^+ \rightarrow RNH_3^+(aq)$$

For example, ethylamine reacts with hydrochloric acid to form a chloride salt:

$$CH_3CH_2NH_2(aq) + HCl(aq) \rightarrow CH_3CH_2NH_3^+(aq) + Cl^-(aq)$$

The formation of amides

Amides can be formed from amines by reaction with carbonyl compounds.

Amides contain the –CONH– functional group. Primary amides have this group at the end of the hydrocarbon chain – for example **ethanamide** is CH_3CONH_2.

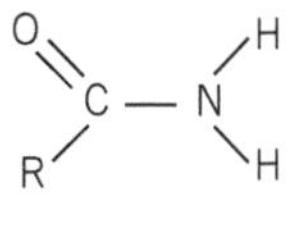

▲ **Figure 1** *A primary amide*

Primary amides have the structure in Figure 1.

Secondary amides have the functional group in the middle of the chain. The nitrogen has one hydrogen and one alkyl group attached to it.

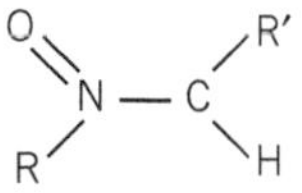

▲ **Figure 2** *A secondary amide*

Secondary amides have the structure in Figure 2.

Revision tip
The NH_2 functional group is called the amino group.

Revision tip
Remember, R stands for an alkyl or aryl group.

Revision tip
You will sometimes see the prefix 'amino' used instead, as in aminobutane for $C_4H_9NH_2$.

Key term
Base: A hydrogen ion (proton) acceptor.

Revision tip
The lone pair on the nitrogen atom forms a dative covalent bond with the hydrogen ion.

Synoptic link
You learned about bases in Topic 8.1, Acids, bases, alkalis, and neutralisation.

Revision tip
The $OH^-(aq)$ ion is responsible for the alkaline characteristics.

Revision tip
The longer the hydrocarbon chain, the less soluble the amine, as hydrocarbon chains are hydrophobic.

Revision tip
When amines dissolve, H_2O acts as an acid, donating a proton to the amine.

Revision tip
Do not confuse am*ine* $(-NH_2)$ with am*ide* (–CONH–).

Revision tip
The carbon in the amide group is counted when determining the name of the amide.

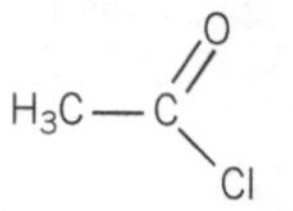

ethanoyl chloride

▲ **Figure 4** *Ethanoyl chloride*

Revision tip
The HCl from these reactions produces white fumes in moist air.

The polymer nylon is an example of a polyamide.

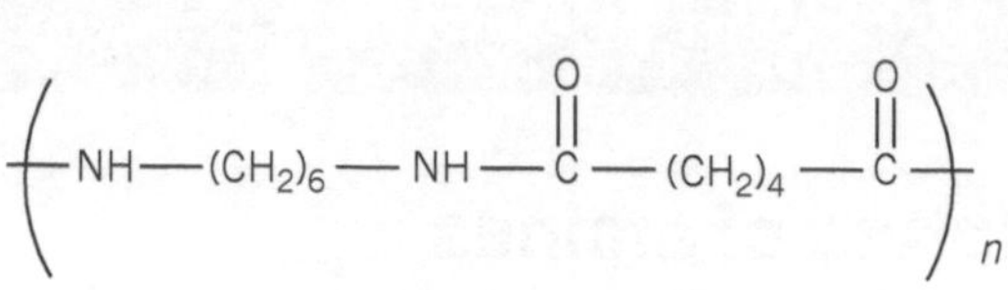

The repeating unit of nylon-6,6

▲ **Figure 3** *The repeating unit of nylon-6,6*

The reactions of amines with acyl chlorides

Acyl chlorides are derivatives of carboxylic acids, with the COCl functional group.

They are reactive forms of carboxylic acids, because the chlorine and oxygen atoms are both significantly more electronegative than carbon.

Acyl chlorides react with ammonia to form primary amides. The general reaction is:

$$RCOCl + NH_3 \rightarrow RCONH_2 + HCl$$

Acyl chlorides react with amines to form secondary amides. The general reaction is:

$$RCOCl + R'NH_2 \rightarrow RCONHR' + HCl$$

For example, ethanoyl chloride (CH_3COCl) reacts with ethylamine as follows:

$$CH_3COCl + C_2H_5NH_2 \rightarrow CH_3CONHC_2H_5 + HCl$$

Acyl chlorides also react with alcohols. In this case, an ester is formed:

$$RCOCl + R'OH \rightarrow RCOOR' + HCl$$

Summary questions

1 Write the structural formulae of:
 a butylamine
 b 2-aminopentane
 c 1,3-diaminopropane. *(3 marks)*

2 Write an equation for the reaction of ethylamine with sulfuric acid. *(2 marks)*

3 Write equations for the reaction of benzoyl chloride, C_6H_5COCl, with:
 a ammonia
 b methylamine
 c methanol. *(3 marks)*

13.7 Hydrolysis of amides and esters

Specification reference: PL(m)

What is hydrolysis?

'Hydro' means water and 'lysis' means to break. Therefore *hydrolysis* involves the breaking of a chemical bond through a reaction with water. Hydrolysis reactions can be considered as the reverse of a condensation reaction.

Hydrolysis of esters

Esters are in equilibrium with their constituent carboxylic acid and alcohol:

$$CH_3CH_2OOCCH_3 + H_2O \rightleftharpoons C_2H_5OH + CH_3COOH$$

Hydrolysis brings about the forward reaction, breaking the C–O bond in the ester and forming the carboxylic acid and alcohol.

Acid hydrolysis of esters

Typically sulfuric acid is used as a catalyst.

Alkaline hydrolysis of esters

In this case, OH^- ions are used as a catalyst. The hydrolysis reaction occurs as shown above, but the carboxylic acid reacts with the OH^- ions, so a **carboxylate salt** is produced instead:

$$CH_3CH_2OOCCH_3 + OH^- \rightarrow C_2H_5OH + CH_3COO^-$$

This has the effect of removing the carboxylic acid from the equilibrium above, and consequently the position of equilibrium moves to the right. Therefore the yield of products increases as the reaction goes to completion.

Hydrolysis of amides

The C–N bond breaks and the products are a carboxylic acid and an amine. The reaction is catalysed by either acid or alkali, resulting in the formation of salts.

Acid hydrolysis of amides

The amide is heated with concentrated sulfuric or hydrochloric acid. The products are a carboxylic acid and the salt of an amine:

$$CH_3CONHCH_3 + H_2O + H^+ \rightarrow CH_3COOH + CH_3NH_3^+$$

Alkaline hydrolysis of amides

The amide is heated with moderately concentrated alkali, typically sodium hydroxide. The products are a carboxylate salt and an amine:

$$CH_3CONHCH_3 + OH^- \rightarrow CH_3COO^- + CH_3NH_2$$

Summary questions

1. Write an equation for the hydrolysis of ethyl ethanoate, $CH_3COOC_2H_5$, to form ethanoic acid and ethanol. *(1 mark)*
2. Write an equation for the alkaline hydrolysis of propyl ethanoate. *(1 mark)*
3. Write equations for the hydrolysis of $CH_3CH_2CONHCH_2CH_3$ under acid and alkaline conditions. *(2 marks)*

Key term

Hydrolysis: The breaking of a C–O bond in an ester or a C–N bond in an amide by reaction with water.

Key term

Carboxylate salt: The product of the reaction of a carboxylic acid with an alkali. A hydrogen ion is lost from the carboxylic acid to leave an $RCOO^-$ ion.

Revision tip

Alkaline hydrolysis of esters is preferred as the yield is better.

Revision tip

The salt is produced because the amine reacts with H^+ ions.

Synoptic link

You learned about the basic nature of amines in Topic 13.6, Amines.

Revision tip

Proteins, which also contain –CONH– groups, can be hydrolysed this way too.

Revision tip

The carboxylate salt is formed because the carboxylic acid reacts with the hydroxide ions.

Synoptic link

You learned about the reactions of carboxylic acids with alkalis in Topic 13.4, Carboxylic acids and phenols, and Topic 13.5, Carboxylic acids.

13.8 Amino acids, peptides, and proteins

Specification reference: PL(a), PL(b), PL(i), PL(q)

The structure of amino acids

Amino acids contain two different functional groups: an amine ($-NH_2$ group), and a carboxylic acid group ($-COOH$ group). Naturally occurring amino acids have these two groups joined to the same carbon atom. This gives amino acids important characteristics:

- They can form **peptide bonds** by condensation polymerisation to make proteins.
- They act as acids *and* as bases.
- They display a type of isomerism known as optical isomerism as they contain **chiral** carbon atoms.

The general formula of an amino acid is $H_2N-CH(R)-COOH$. The R-group is a side chain, and each amino acid has a unique R-group. Some examples are shown in Table 1.

▼ **Table 1** *Some amino acids*

R-group	Name and abbreviation of amino acid	Structure
H	Glycine, Gly	H_2NCHCO_2H $\|$ H
CH_3	Alanine, Ala	H_2NCHCO_2H $\|$ CH_3
CH_2OH	Serine, Ser	H_2NCHCO_2H $\|$ CH_2OH
CH_2SH	Cysteine, Cys	H_2NCHCO_2H $\|$ CH_2SH
$CH(CH_3)_2$	Valine, Val	H_2NCHCO_2H $\|$ $CHCH_3$ $\|$ CH_3

There are 20 different naturally occurring amino acids, all with unique R-groups.

Optical isomerism

All amino acids, except glycine, have four different groups attached to the central carbon atom. This gives rise to **optical isomers**.

Chirality

Amino acids, except glycine, contain a chiral centre. Four groups can be arranged in different ways around a carbon atom. This is a type of stereoisomerism, and it is known as optical isomerism.

Figure 1 shows the two optical isomers of alanine. They are mirror images of each other, but the mirror images are non-superimposable.

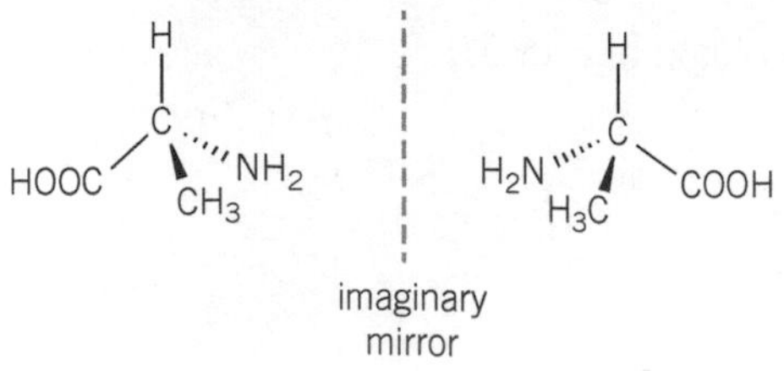

imaginary mirror

▲ **Figure 1** *The optical isomers of alanine*

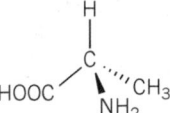

If you rotate the right-hand image you can see that it is not the same as the left-hand image, see Figure 2. The $-CH_3$ group and the $-NH_2$ group are not in the same position. They are separate molecules. The only way to make them superimpose would be to break the bonds, but this would require a chemical reaction to make the new molecule.

Whenever a molecule contains a chiral carbon, with four different groups attached, it will have optical isomers. The optical isomers are sometimes known as **enantiomers**. Enantiomers are separate molecules to each other, although most of their physical properties, such as melting point, are the same. However their biological activity can be very different, due to the way they may fit into the active site of an enzyme, and they sometimes smell or taste different to each other.

Acid–base chemistry of amino acids

In alkaline conditions, the $-COOH$ group of an amino acid can lose a proton, leaving a $-COO^-$ group:

$$H_2N-CHR-COOH \rightarrow H_2N-CHR-COO^- + H^+$$

In acidic conditions, the $-NH_2$ group can accept a proton as the nitrogen has a lone pair and can form a dative covalent bond with H^+:

$$H_2N-CHR-COOH + H^+ \rightarrow H_3N^+-CHR-COOH$$

Amino acids can also exist as zwitterions, where the NH_2 group is protonated and the $COOH$ is deprotonated. This happens in neutral solution.

Amino acids are soluble in water due to their ionic nature. Adding small quantities of acid or alkali does not have much effect on the pH as the zwitterions act like buffers, accepting or donating H^+ ions.

The ionic forms of alanine are shown below.

In acid solution	In neutral solution	In alkaline solution
$H_3N^+-CH(CH_3)-COOH$	$H_3N^+-CH(CH_3)-COO^-$	$H_2N-CH(CH_3)-COO^-$
NH_2 group is protonated due to high concentration of H^+.	Zwitterion.	COOH group deprotonated due to high concentration of OH^-.

Proteins – condensation polymers of amino acids

Amino acids are bifunctional compounds, and the $-COOH$ group can react with the $-NH_2$ group of a neighbouring molecule in a condensation reaction:

$$H_2N-CHR-COOH + H_2N-CHR'-COOH \rightarrow H_2N-CHR-CONH-CHR'-COOH + H_2O$$

The $-CONH-$ group is known as a **peptide link**, and the product above is known as a **dipeptide**. Figure 3 shows the dipeptide formed from glycine and alanine.

A polypeptide, or protein, is a molecule containing many amino acid residues joined together with peptide links.

The primary structure of proteins

Insulin is a protein made from 51 amino acid monomers. The precise order in which the amino acids are joined together is known as the **primary structure** of the protein. For example, the primary structure of insulin begins Gly-Ile-Val-Glu-Gln-Cys…, where each three-letter code refers to an amino acid.

All proteins have different primary structures as they are made of amino acids joined together in different orders.

▲ **Figure 2** *Rotating the molecule*

Key term

Enantiomer: Either one of a pair of optical isomers.

Synoptic link

The acid–base chemistry of amines and carboxylic acids are covered in Topic 13.5, Carboxylic acids and 13.6, Amines.

Key term

Zwitterion: An amino acid in which the NH_2 group is protonated to form NH_3^+, and the COOH group is deprotonated to form COO^-.

▲ **Figure 3** *A peptide link between glycine and alanine*

Revision tip
A peptide link is a secondary amide.

Synoptic link

You learned about amides in Topic 13.7, Hydrolysis of esters and amides, and 13.8, Amino acids, peptides, and proteins.

Revision tip

A protein is said to contain amino acid 'residues' as the amino acid lost the elements of water when the peptide bond was formed.

Hydrolysis of peptide links

Peptides are secondary amides, and hydrolysis of the C–N bond produces the amino acids that the protein is composed of. The protein is hydrolysed by heating with moderated concentrated acid or alkali.

Paper chromatography

Paper chromatography can be used to identify the amino acids that are present in a protein. Once the protein has been hydrolysed, a spot of the product mixture can be placed on a piece of chromatography paper.

Using a suitable solvent, the chromatogram is allowed to run, and the different amino acids in the mixture will separate as the solvent rises up the paper. To reveal the spots, a locating agent, or ultraviolet light, may be needed.

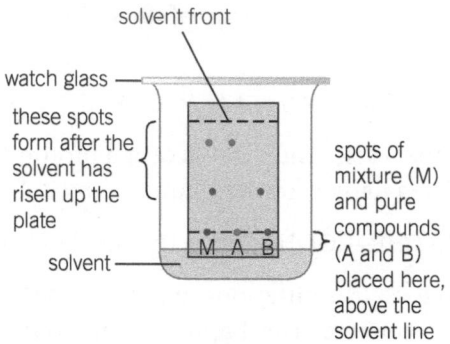

▲ **Figure 4** *Paper chromatography*

R_f values can be calculated: $Rf = \dfrac{\text{distance moved by spot}}{\text{distance moved by the solvent}}$. R_f values are always between 0 and 1, and have no units as the units of distance cancel out.

By comparing the R_f values with known data, or by repeating the chromatography with known samples of amino acids under the same conditions, the identity of the amino acids from the hydrolysed protein can be determined.

Summary questions

1. Draw the structure of valine under:
   a alkaline conditions
   b neutral conditions
   c acidic conditions. *(3 marks)*

2. Draw the two different dipeptides that can be formed by serine and alanine. *(2 marks)*

3. Draw the products of the hydrolysis of the dipeptide $H_2NCH_2CONHCH(CH_3)COOH$ when hydrolysed with:
   a hydrochloric acid
   b sodium hydroxide. *(2 marks)*

13.9 Oils and fats

Specification reference: CD(c)

Oils and fats are naturally occurring **triesters** of propane-1,2,3-triol (known as glycerol) and long-chain carboxylic acids (known as fatty acids). The three ester groups can be identical, from the same fatty acid, or they can be different. Naturally occurring oils and fats tend to have different fatty acid groups.

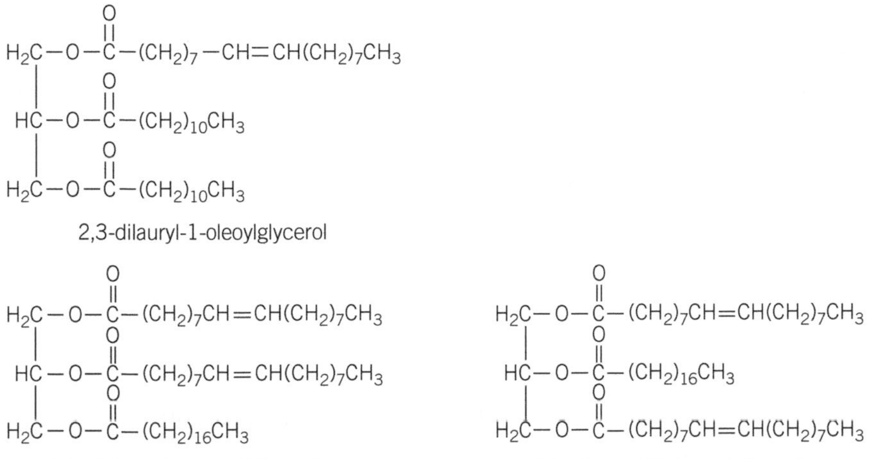

$$H_2C-O-\overset{\overset{O}{\|}}{C}-(CH_2)_7-CH=CH(CH_2)_7CH_3$$
$$HC-O-\overset{\overset{O}{\|}}{C}-(CH_2)_{10}CH_3$$
$$H_2C-O-\overset{\overset{O}{\|}}{C}-(CH_2)_{10}CH_3$$

2,3-dilauryl-1-oleoylglycerol

$$H_2C-O-\overset{\overset{O}{\|}}{C}-(CH_2)_7CH=CH(CH_2)_7CH_3$$
$$HC-O-\overset{\overset{O}{\|}}{C}-(CH_2)_7CH=CH(CH_2)_7CH_3$$
$$H_2C-O-\overset{\overset{O}{\|}}{C}-(CH_2)_{16}CH_3$$

1,2-dioleoyl-3-stearoylglycerol

$$H_2C-O-\overset{\overset{O}{\|}}{C}-(CH_2)_7CH=CH(CH_2)_7CH_3$$
$$HC-O-\overset{\overset{O}{\|}}{C}-(CH_2)_{16}CH_3$$
$$H_2C-O-\overset{\overset{O}{\|}}{C}-(CH_2)_7CH=CH(CH_2)_7CH_3$$

1,3-dioleoyl-2-stearoylglycerol

▲ **Figure 1** *Three different triesters*

Fatty acids

The carboxylic acids in oils and fats are known as 'fatty acids'. They are long hydrocarbon chains, and are usually unbranched. They have an even number of carbon atoms in their chains, up to 24, but typically contain 16 or 18. The presence of these long hydrocarbon chains, which are non-polar, means that oils and fats do not mix with water.

The fatty acid chains can be saturated, containing only single C–C bonds, or unsaturated, with one or more C=C double bonds.

Unsaturated fatty acid chains can be hydrogenated to make the fat more spreadable. In this process C=C bonds react with H_2 to produce C–C bonds.

Hydrolysis of triesters

Hydrolysing a triester involves heating it with concentrated acid or alkali. This produces the three fatty acids and propane-1,2,3-triol.

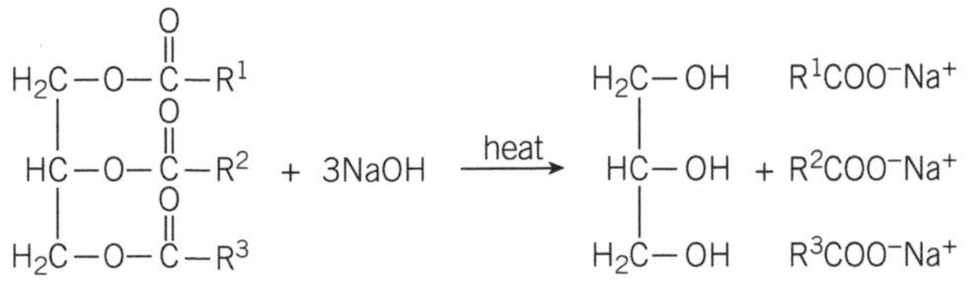

$$H_2C-O-\overset{\overset{O}{\|}}{C}-R^1$$
$$HC-O-\overset{\overset{O}{\|}}{C}-R^2 + 3NaOH \xrightarrow{heat}$$
$$H_2C-O-\overset{\overset{O}{\|}}{C}-R^3$$

$$H_2C-OH \quad R^1COO^-Na^+$$
$$HC-OH + R^2COO^-Na^+$$
$$H_2C-OH \quad R^3COO^-Na^+$$

▲ **Figure 2** *The hydrolysis of a triester*

Summary questions

1 What is meant by the term 'mixed triester'? *(1 mark)*

2 Draw the products of hydrolysis of a triester under alkaline conditions. Use R^1, R^2, and R^3 to represent the fatty acid chains. *(1 mark)*

3 A mixed triester was hydrogenated. 0.15 mol of the triester required 7.2 dm³ of hydrogen at room temperature and pressure for complete hydrogenation. How many C=C bonds did the triester contain? *(3 marks)*

Synoptic link

You learned about esters in Topic 13.3, Alcohols, and Topic 13.4, Carboxylic acids and phenols.

Revision tip

The only difference between oils and fats is that oils are liquid at room temperature, and fats are solid.

Revision tip

Triesters are sometimes called **triglycerides** because they are based on glycerol.

Key term

Fatty acid: A long-chain carboxylic acid, found as an ester in an oil or fat.

Revision tip

Unsaturated fats are considered healthier than saturated fats.

Revision tip

Hydrolysing an oil or fat with an alkali tends to make the reaction go to completion (see Topic 13.7, Hydrolysis of esters and amides).

Revision tip

Hydrolysis of a triester with NaOH produces sodium salts of the fatty acids. The free fatty acids can be released by treating the sodium salt with dilute HCl.

Revision tip

Remember that three moles of NaOH are required to hydrolyse a triester.

13.10 Aldehydes and ketones

Carbonyl compounds

Many functional groups, such as carboxylic acids and amides, contain C=O bonds. This section considers two homologous series containing the carbonyl group, C=O, aldehydes and ketones.

Aldehydes and ketones

Aldehydes have a carbonyl group at the end of a carbon chain. They have the general formula R–CHO. Their names end in –al.

Ketones have a carbonyl group in the middle of a chain. They have the general formula R^1–CO–R^2. Their names end in –one.

Although they contain the same functional group, aldehydes are more reactive than ketones as they can be oxidised to form carboxylic acids.

Oxidation of aldehydes

Aldehydes can be oxidised using acidified potassium dichromate(VI) to form the corresponding carboxylic acid:

$$RCHO + [O] \rightarrow RCOOH$$

For example, methanal reacts to form methanoic acid:

$$CH_2O + [O] \rightarrow HCOOH$$

Ketones **cannot** be oxidised because they do not contain a hydrogen attached to the carbonyl group.

Distinguishing between aldehydes and ketones

Because aldehydes can be oxidised, they can be distinguished from ketones by refluxing with acidified potassium dichromate. Aldehydes cause a colour change from orange to green, whereas the colour remains orange with ketones.

Fehling's solution

The test substance is warmed with Fehling's solution. Fehling's solution contains Cu^{2+} ions, which oxidise aldehydes. This causes the Cu^{2+} ions to be reduced and a precipitate of copper(I) oxide, Cu_2O, is formed. The colour changes from blue to red. This is a positive test for aldehydes. Ketones do not cause the colour change.

Tollens' reagent

Tollens' reagent contains Ag^+ ions, which oxidise aldehydes. When an aldehyde is warmed with Tollens' reagent, the Ag^+ ions are reduced to metallic Ag, which appears as a silvery layer on the inside of the test tube. Ketones do not produce a silver mirror.

Reaction with hydrogen cyanide, HCN

Aldehydes and ketones can react with HCN to produce a cyanohydrin. This is a nucleophilic addition reaction.

Hydrogen cyanide is too hazardous to use in the laboratory, so acidified KCN is often used instead.

Revision tip

The aldehyde functional group is written as CHO, not COH, in order to distinguish it from alcohols.

Synoptic link

In Topic 13.3, Alcohols, you learned that aldehydes are produced by the oxidation of primary alcohols, and ketones are produced by the oxidation of secondary alcohols.

Revision tip

[O] represents an oxygen from the oxidising agent.

Revision tip

Because acidified dichromate solution also oxidises alcohols, this is not a conclusive test for aldehydes.

Key term

Cyanohydrin: A molecule containing an OH and a CN group.

Revision tip

The CN^- ion has a lone pair which acts as a nucleophile, to form a covalent bond.

Revision tip

This is an addition reaction because the HCN molecule is added across the C=O bond.

Nucleophilic addition

- The cyanide ion is attracted to the partial positive charge of the carbon atom in the C=O bond.
- A new carbon–carbon bond is formed.
- A pair of electrons from the C=O bond moves to the oxygen atom, which then becomes negatively charged.
- The negatively charged ion then picks up a hydrogen ion, H^+, from the solvent water.

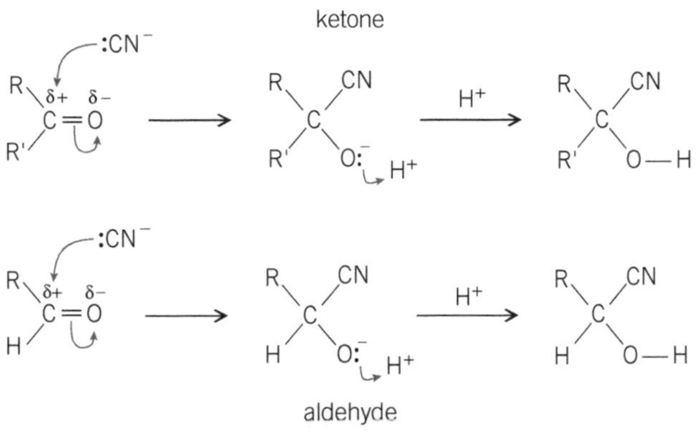

▲ **Figure 1** *The mechanism of nucleophilic addition of a ketone (top) and an aldehyde (bottom)*

This sort of reaction is valuable in organic synthesis as it creates a new carbon–carbon bond. Therefore it is a way of increasing the carbon chain length.

> **Revision tip**
> The C=O bond is polarised because oxygen is more electronegative than carbon.

> **Revision tip**
> Remember, a curly arrow represents the movement of a pair of electrons.

Summary questions

1 A student has mixed up two bottles, one of which contains an aldehyde and one of which contains a ketone. Suggest how the student could identify them using a chemical test. *(2 marks)*

2 Write an equation for the oxidation of :
   a propanal
   b propanone. *(2 marks)*

3 Draw the mechanism of the nucleophilic addition of H–CN to ethanal. *(3 marks)*

Chapter 13 Experimental techniques

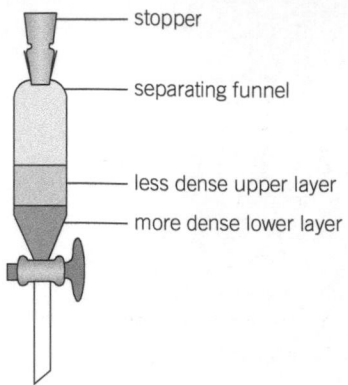

stopper

separating funnel

less dense upper layer

more dense lower layer

▲ **Figure 1** *A separating funnel*

Revision tip

The lower layer is the denser liquid.

Revision tip

If you are not sure which layer is the aqueous layer, you can add a few drops of water to the separating funnel and watch to see which layer they join.

Purifying an organic liquid product

1 Two immiscible liquids can be separated using a separating funnel. Often an organic product (hydrophobic) is mixed with an aqueous solution. The two layers will separate and can be run off through the seaparating funnel. The aqueous layer can be disposed of.

2 Acidic impurities can be neutralised by adding sodium hydrogen carbonate solution. Alkaline impurities can be neutralised by adding a dilute acid.

3 The crude product can be dried using an anhydrous salt such as anhydrous sodium sulfate or calcium chloride.

4 The pure product can be separated by distillation.

Recrystallisation

Recrystallisation is used to purify solid organic products, by dissolving them in a hot solvent. When cooled, the pure compound will form (recrystallise), with soluble impurities remaining in solution.

1 A suitable solvent is one in which the organic product is very soluble at higher temperatures and insoluble (or virtually insoluble) at lower temperatures.

2 Dissolve the impure solid in the minimum quantity of hot solvent.

3 Filter to remove impurities.

4 Leave the filtrate to cool until crystals form.

5 Collect the crystals by vacuum filtration, and dry them in an oven or in the open, covered by an upside-down filter funnel.

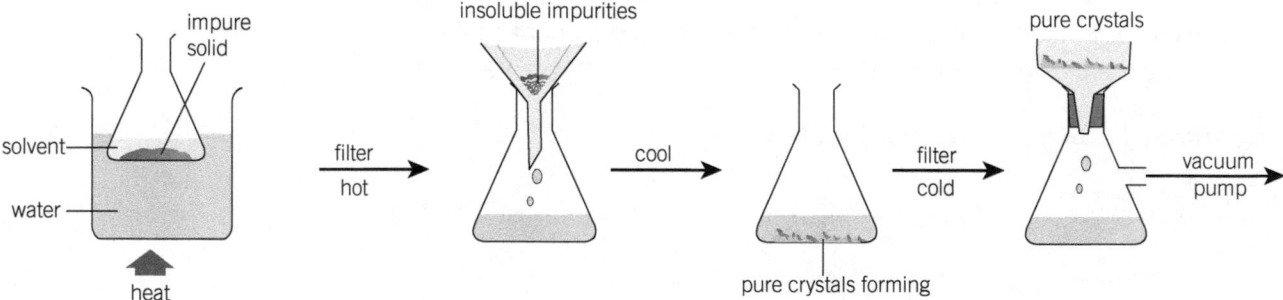

▲ **Figure 2** *Recrystallisation of an impure solid*

Revision tip

You will need to remove the stopper when you are running the liquids off from the separating funnel.

Revision tip

It is important to use the minimum quantity of solvent, as this will give a better yield of product.

Revision tip

It is best to preheat the filter funnel and conical flask to prevent early recrystallisation.

Vacuum filtration

Vacuum filtration enables you to separate a solid rapidly from a filtrate.

1 Connect a side-arm conical flask to a vacuum pump.

2 Place a damp piece of filter paper in the Buchner funnel. The paper should be placed flat.

3 Switch on the vacuum pump and carefully pour in the mixture to be filtered. The partial vacuum ensures that the filtration happens quickly.

Determining melting points

The melting point of an organic solid is evidence of the identity of the substance and of its purity.

1 Heat a glass melting point tube in a Bunsen flame to seal the end.

2 Put a small amount of the solid into the tube by tapping the open end of the tube into the solid, so that a small quantity goes in. Tap the tube so that the solid falls to the bottom of the sealed end.

3 Place the tube in the melting point apparatus and begin heating.

4 Note the temperature range over which the solid melts – when it starts melting, and when it finishes melting.

Thin-layer and paper chromatography

Chromatography involves a mobile phase (a solvent) and a stationary phase (paper or, in thin-layer chromatography, a silica plate). It relies on the fact that different compounds have different affinities for different mobile phases, and so are carried up the stationary phase at different rates. Therefore chromatography can be used to identify the components of a mixture.

Paper chromatography and thin-layer chromatography (TLC) use very similar procedures:

1 Draw a pencil line 1 cm from the base of the chromatography paper or plate.

2 Spot the test mixture and reference samples along the line.

3 Suspend the paper or plate in a covered beaker containing the solvent.

4 Wait for the solvent to reach near the top. Remove the plate, mark the solvent front, and allow it to dry.

5 Locate the spots using iodine, ninhydrin, or ultraviolet light.

6 Match the heights reached, or R_f values, with known data using the same solvent.

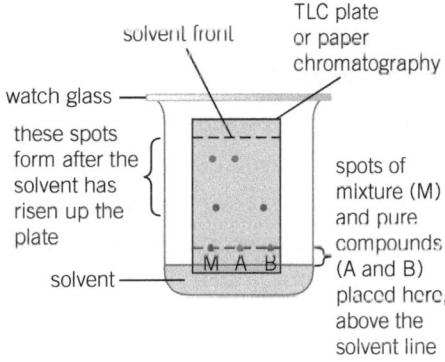

solvent front

TLC plate or paper chromatography

watch glass

these spots form after the solvent has risen up the plate

solvent

spots of mixture (M) and pure compounds (A and B) placed here, above the solvent line

M A B

▲ **Figure 3** *Thin-layer or paper chromatography*

> **Revision tip**
> The experimental melting point range can be compared with published values.

> **Revision tip**
> A pure compound will melt within 0.5°C of the true melting point.

> **Revision tip**
> An impure compound will have a wider melting range.

Chapter 13 Practice questions

1 Which statement is true about carboxylic acids? *(1 mark)*

   A Carboxylic acids react with metals to produce carbon dioxide.

   B The formula of calcium ethanoate is CH_3COOCa.

   C Carboxylic acids are strong acids.

   D Carboxylic acids react with alcohols to form esters.

2 Which of the following reactions of amines is **incorrect**? *(1 mark)*

   A $CH_3NH_2 + H^+ \rightarrow CH_3NH_3^+$

   B $CH_3NH_2 + OH^- \rightarrow CH_3NH^- + H_2O$

   C $CH_3NH_2 + HCl \rightarrow CH_3NH_3^+ Cl^-$

   D $CH_3NH_2 + CH_3COCl \rightarrow CH_3NHOCCH_3 + HCl$

3 What are the products of alkaline hydrolysis of $C_6H_5COOC_2H_5$? *(1 mark)*

   A $C_6H_5OH + C_2H_5COO^-$ **C** $C_6H_5COO^- + C_2H_5OH$

   B $C_6H_5O^- + C_2H_5COOH$ **D** $C_6H_5COOH + C_2H_5O^-$

4 Which statement describes a property of the amino acid alanine, $H_2NCH(CH_3)COOH$? *(1 mark)*

   A Alanine can form two different dipeptides with glycine, H_2NCH_2COOH.

   B Alanine is insoluble in water.

   C Alanine has no isomers.

   D Alanine is deprotonated at low pH.

5 Pentanoic acid occurs naturally in the flowering plant valerian. It has a distinctive, unpleasant odour.

   a Give the molecular formula of pentanoic acid. *(1 mark)*

   b Write an equation for the reaction of pentanoic acid with sodium carbonate. Explain how this is used as a test for carboxylic acids. *(2 marks)*

   c Pentanoic acid reacts with ammonia. Write an equation for this reaction and classify the product as a primary or secondary amide. *(2 marks)*

6 The tripeptide Gly-Ala-Ser was hydrolysed by a student.

   a Draw the structure of the tripeptide before hydrolysis. (Assume that the NH_2 group has not formed a peptide link and is the 'N terminal'.) *(1 mark)*

   b Explain why different products are produced, depending on whether acid or alkaline hydrolysis was used. *(1 mark)*

   c Explain why only two of the products display optical isomerism. Identify the product that does not display optical isomerism. *(3 marks)*

   d Describe how paper chromatography could be used to identify the products of the hydrolysis. *(3 marks)*

7 The naturally occurring compounds heptan-2-one, $C_7H_{14}O$, found in blue cheese, and benzaldehyde, C_6H_5CHO, found in almonds, are both colourless liquids at room temperature. A student carried out a range of reactions on both substances.

   a Which homologous series do these compounds belong to? *(1 mark)*

   b Draw the structure of 2-heptanone. *(1 mark)*

   c Which of the compounds will react with Fehling's solution? What will be observed during the reaction? *(2 marks)*

   d Which of the compounds will react with Tollens' reagent? What will be observed during the reaction? *(2 marks)*

   e Which of the compounds will react with HCN? What product(s) will be formed? *(3 marks)*

14.3 Operation of a chemical manufacturing process

Specification reference: CI (i), CI(k)

Chemical reactions in industry

Industry uses a wide range of reaction types. Examples you encounter in this course include:

- Direct combination of elements to make simple molecules

  $N_2 + 3H_2 \rightleftharpoons 2NH_3$

- Simple oxidation processes

  $2SO_2 + O_2 \rightarrow 2SO_3$

- Reduction of the oxides in metal ores

  $Fe_2O_3 + 3CO \rightarrow 2Fe + 3CO$

- Substitution reactions involving organic molecules

  $C_6H_6 + HNO_3 \rightarrow C_6H_5NO_2 + H_2O$

- Polymerisation reactions

  $nCH_2CH_2 \rightarrow (CH_2CH_2)_n$

Choices in chemical industry

Chemists need to take decisions about how to manufacture a particular product. These decisions are taken to attempt to ensure that the process is as cost-effective as possible whilst minimising the environmental impact.

Factors that need to be considered include:

- Choice of starting materials (feedstocks) for the reaction
- Conditions needed for the reaction (pressure, temperature)
- The possible use of a catalyst
- The atom economy of the reaction used
- The yield of the reaction
- The rate at which products are formed
- The formation of any waste products that need to be disposed of
- The hazards associated with feedstocks or products.

Most of these factors are described in more detail in other chapters of this guide.

Costs of industrial processes

There are several components to the cost of an industrial process:

- The cost of the raw materials and any costs associated with converting them into feedstocks that can be added to the reactor.

  For example, crude oil is a relatively cheap raw material, although there are considerable costs associated with the fractional distillation and cracking needed to extract ethene molecules which are necessary for polymerisation reactions.

- Energy costs. Fuel needs to be burnt, or electricity used to create high temperatures and pressures.

- Costs associated with plant. Many chemical processes require expensive specialist apparatus – for example thick-walled reactors to enable reactions to take place at high pressures.

> **Revision tip**
>
> You will not be required to recall the details of particular industrial processes, but if you are given information about a process you will need to be prepared to use and analyse the information provided.

> **Key term**
>
> **Raw materials:** The naturally occurring resource from which the feedstock is obtained. Examples include crude oil, natural gas, metal ores, the air, sea water, and biomass such as plant material.

> **Synoptic link**
>
> The factors affecting the rate of reaction are described and explained in Topic 10.1, Factors affecting reaction rates, and Topic 10.2, The effect of temperature on rate.

> **Synoptic link**
>
> The factors affecting equilibrium position of industrial reactions are covered in Topic 7.5, Equilibrium constant K_c, temperature, and pressure.

> **Synoptic link**
>
> Atom economy is explained in Topic 14.2, Atom economy.

Revision tip

If details of pressure and temperature are provided, you should discuss the effect that these conditions will have on both the yield and the rate of the reaction.

Discussing the conditions used in the operation of a chemical process

An example of how to do this is provided in the model answer below:

Model answer: Manufacture of ammonia

Ammonia, NH_3 is manufactured from nitrogen and hydrogen. The conditions used are 200 atmospheres and 450 °C with the use of an iron catalyst. Discuss why these are the most cost-effective conditions to use.

High pressure increases the rate of the reaction and the yield at equilibrium (because the equilibrium position favours the side of the equation with the fewest moles). However, high pressure requires a lot of energy to maintain and requires expensive reaction vessels.

> The answer discusses the effect of the high pressure on rate and yield.

High temperature increases the rate but decreases the yield (because the equilibrium position favours the endothermic direction). However, high temperature requires a lot of energy to maintain.

> The answer discusses the effect of the moderately high temperature on rate and yield.

A very high pressure is chosen, but above 200 atmospheres the extra costs outweigh the increase in rate and yield. A moderately high temperature is a compromise, to produce an acceptable yield without decreasing the rate too much. The use of a catalyst helps to increase the rate without the need for high temperature.

> The answer explains how the chosen conditions are the most cost effective, using ideas from the previous two paragraphs.

Co-products and by-products

The overall cost of operating an industrial process is also affected by the formation of any additional products formed alongside the main product. Additional products will need to be separated from the main product which increases the cost of the process. They may also be hazardous or require special handling. However, if the additional products can be sold then this can help to make the overall process more profitable.

Key terms

Co-product: An additional product formed as a result of the intended reaction in a chemical process.

By-product: An additional product formed as a result of a unintended reaction.

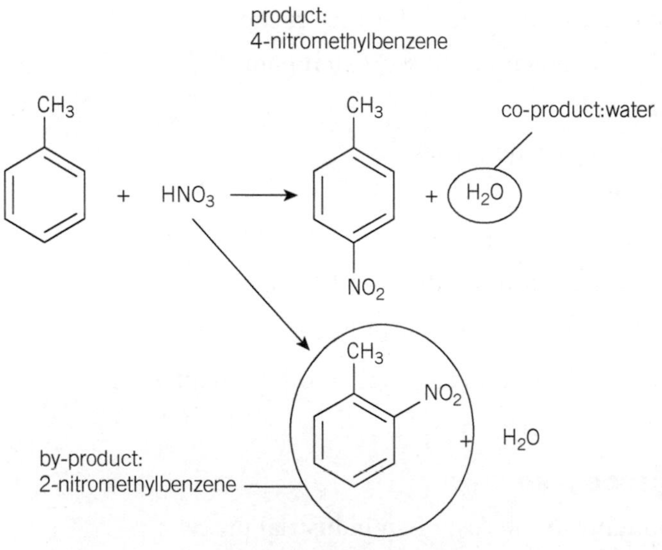

▲ **Figure 1** *In the nitration of methylbenzene, water is a co-product because it is formed alongside the product in the intended reaction. 2-nitromethylbenzene is a by-product because it is formed in an unintended reaction*

Choices for society

If a product is particularly hazardous, then society may need to take an ethical decision about whether its manufacture and use should be permitted.

These decisions may be based on:

- the benefit that the product may bring to individuals or society.

This is weighed up against:

- the risks or hazards associated with the use or manufacture.

An example of how to do this is provided in the model answer.

Revision tip

You may be required to discuss the advantages and disadvantages of the use and manufacture of a particular product, and reach a reasoned conclusion about whether its use and manufacture is justified.

Model answer: Discuss the risks and benefits in the use of chlorine in water treatment

Chlorine kills bacteria that may cause serious or fatal diseases such as cholera.

However chlorine can be harmful to humans if the concentration in water is too high.

There are also risks associated with leaks of chlorine gas during manufacture or transport.

If suitable safety precautions are taken during transport, and if the concentration of the chlorine is kept to as low a level as possible, then the benefits of the use and manufacture of chlorine outweigh the risks.

> Describe a benefit of using chlorine.

> Describe some of the hazards and risks of handling or using chlorine.

> Balance the benefits against the risks and reach a justified conclusion about the use if chlorine.

Summary questions

1 2-chlorobutane can be converted to but-2-ene in an elimination reaction:

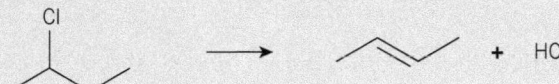

A small amount of but-1-ene is also formed.
Classify HCl and but-1-ene as either by-product or co-product and explain your answer. *(3 marks)*

2 Many important products can be prepared in two (or more) ways. For this pair of processes, select the process which is likely to be the **greenest** (has the lowest environmental impact). Give a reason for your answer.
Chloroethane can be synthesised in industry by two different processes:
 A From the reaction of ethene and hydrogen chloride at 130 °C with an aluminium chloride catalyst
 $C_2H_4 + HCl \rightarrow C_2H_5Cl$
 B From the reaction of ethanol and hydrochloric acid over a zinc chloride / aluminium oxide catalyst at 325 °C
 $C_2H_5OH + HCl \rightarrow C_2H_5Cl + H_2O$
 Suggest and explain three reasons why reaction A is currently more cost effective than reaction B *(6 marks)*

3 Hydrcarbons, such as methane, can be formed from CO and water. The exothermic reaction is carried out at 400 °C and atmospheric pressure, using a catalyst that is a mixture of iron(III) oxide and chromium(III) oxide:
 $CO + H_2O \rightleftharpoons CO_2 + H_2$
 a Explain the reasons for the choice of these conditions. *(7 marks)*
 b Discuss whether the use of this reaction will be an overall benefit to society, bearing in mind the risks and benefits of the substances involved. *(5 marks)*

1 Nitrogen oxide is converted into nitrogen dioxide in industry:

$2NO(g) + O_2(g) \rightleftharpoons 2NO_2(g)$ $\Delta_rH = -115\,kJ\,mol^{-1}$

Which changes will produce an increase in both rate and yield?

A Using a catalyst and increasing the pressure.

B Increasing the temperature and decreasing the pressure.

C Using a catalyst and increasing the temperature.

D Decreasing the pressure and decreasing the temperature.   (*1 mark*)

2 Titanium(IV) oxide can be formed from titanium(IV) chloride by oxidation at high temperature. $TiCl_4(l) + O_2(g) \rightarrow TiO_2(s) + 2Cl_2(g)$

Which aspect is likely to be a significant environmental issue?

1 The reaction must be carried at high pressure, increasing the risk of explosions.

2 The raw materials used to produce oxygen are not renewable.

3 The atom economy of the process is very low.

A 1,2, and 3   **B** 1 and 2   **C** 2 and 3   **D** Only 3   (*1 mark*)

3 Nylon 6,6 can be formed in the laboratory from hexanedioyl dichloride and 1,6-diaminohexane. However in industry hexanedioic acid is used in place of hexanedioyl dichloride. The main reason for this is that:

A Hexanedioyl dichloride must be formed from crude oil which is non-renewable.

B Using hexanedioyl dichloride creates a toxic co-product during the reaction.

C Hexanedioic acid reacts more rapidly than hexanedioyl dichloride.

D No catalyst is necessary when using hexanedioic acid.   (*1 mark*)

4 Which of these types of reaction will have the highest atom economy?

A Substitution   **B** Elimination   **C** Addition   **D** Condensation (*1 mark*)

5 Ammonium nitrate can be made from the reaction of ammonia and concentrated nitric acid: $NH_3(g) + HNO_3(aq) \rightarrow NH_4NO_3(aq)$

Which of these are likely to be significant hazards of this process:

1 Ammonia is flammable.

2 Nitric acid is corrosive.

3 The reaction is likely to be highly exothermic.

A 1,2, and 3   **B** 1 and 2   **C** 2 and 3   **D** only 3   (*1 mark*)

6 As part of the process that results in the manufacture of nitric acid, ammonia is oxidised to nitrogen oxide (NO).

The equation for the reaction is:

$4NH_3(g) + 5O_2(g) \rightleftharpoons 4NO(g) + 6H_2O(g)$ $\Delta_rH = -905.2\,kJ\,mol^{-1}$

Under some conditions, some nitrogen dioxide, NO_2 may also be formed.

a Calculate the atom economy of this process, and comment on your answer.   (*2 marks*)

b Identify a **co-product** and a **by-product** of the reaction to form nitrogen oxide. Explain your answer.   (*4 marks*)

i The reaction is carried out industrially at a temperature of 500 K. Explain why a higher temperature is not used.   (*2 marks*)

ii A pressure of about 5 atmospheres is often chosen for the reaction. Discuss the reasons for the choice of 5 atmospheres.   (*6 marks*)

15.3 Radiation in, radiation out

Specification reference: O(n)

The greenhouse effect

The temperature of the Earth's surface depends on the balance between energy absorbed from the Sun and energy emitted by the Earth's surface, which is then lost into space. If there were no atmosphere, the average temperature of the Earth's surface would be about 254 K (−19 °C).

The presence of greenhouse gases in the troposphere causes a greenhouse effect that increases the average temperature of the Earth's surface to about 287 K (14 °C).

Human activities have increased the concentration of greenhouse gases in the troposphere. This is resulting in an enhanced greenhouse effect, raising temperatures even higher. Current models suggest that unless emissions of greenhouse gases are reduced significantly, temperatures may rise a further 3 °C or 4 °C higher by the end of the century.

Emissions from the Sun and the Earth

The Sun and the Earth both emit electromagnetic radiation. However, because of the different temperatures at the surface of these bodies, they emit different types of electromagnetic radiation.

▼ **Table 1** *Radiation emitted from the Sun and the Earth*

Body	Surface temperature	Main type(s) of radiation emitted	Approximate range of frequencies emitted	Approximate range of wavelengths emitted
Sun	6000 K	Ultraviolet Visible Some infrared	10^{14}–10^{15} Hz	3000 to 300 nm
Earth	287 K	Infrared	10^{12}–10^{13} Hz	188 30 000 nm to 300 000 nm

Effect on molecules in the Earth's atmosphere

- Ultraviolet radiation can be absorbed by molecules such as ozone in the atmosphere and cause bond breaking.

- Most of the visible radiation from the Sun passes through the atmosphere without being absorbed.

- Infrared radiation from the Earth can be absorbed by the bonds in molecules, causing them to vibrate with greater energy.

The infrared window

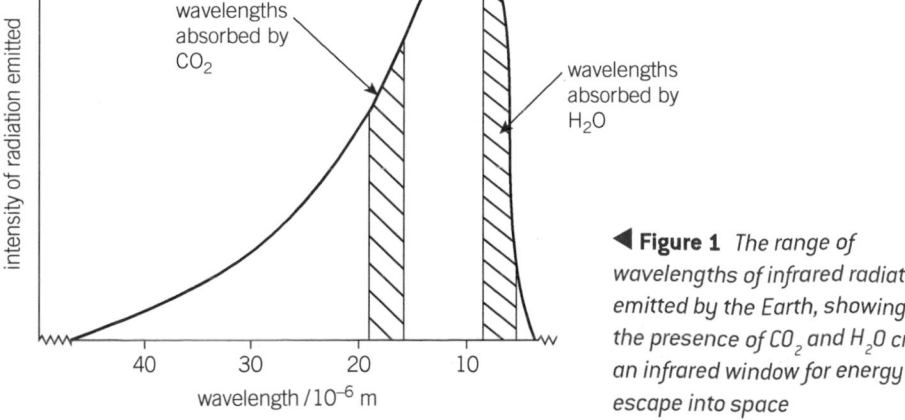

◄ **Figure 1** *The range of wavelengths of infrared radiation emitted by the Earth, showing how the presence of CO_2 and H_2O create an infrared window for energy to escape into space*

Key terms

Greenhouse effect: The process in which infrared radiation emitted by the Earth is absorbed by gases in the atmosphere, causing a temperature increase.

Greenhouse gases: Gases that absorb infrared radiation emitted by the Earth.

Troposphere: The lowest layer of the Earth's atmosphere.

Revision tip

You will not be expected to recall the frequency and wavelength ranges of the radiation emitted by the Sun and the Earth, although you do need to know the names of the types of radiation involved.

You may also need to do calculations to convert between frequency and wavelength.

Synoptic link

The electromagnetic spectrum was introduced in Topic 6.2, What happens when radiation interacts with matter?

Synoptic link

You can remind yourself how to convert between frequency and wavelength using the equation $c = f \times \lambda$ in Topic 6.1, Light and electrons.

Synoptic link

Bond breaking caused by ultraviolet radiation is described in Topic 6.3, Radiation and radicals, and the effect of infrared radiation on bonds in molecules is described in Topic 6.4, Infrared spectroscopy.

The most abundant greenhouse gas in the troposphere is water vapour. This absorbs frequencies of infrared radiation that fall within the range of wavelengths emitted by the Earth's surface, leaving a 'window' of infrared wavelengths that can pass through.

Increasing the concentration of CO_2 or other greenhouse gases will increase the amount of total infrared radiation absorbed, and decrease the amount of radiation escaping into space.

Other greenhouse gases absorbing radiation in the 'window' include any molecules containing C–F bonds, for example chlorofluorocarbons (CFCs).

Greenhouse gases

▼ **Table 2** *Main sources of greenhouse gases*

Gas	Main source from human activity
carbon dioxide	burning of fossil fuels
methane	cattle farming
dinitrogen oxide	decomposition of nitrogen-based fertilisers
haloalkanes	previously used as refrigerants, solvents etc.

* Although water vapour is an abundant greenhouse gas, the emissions of water vapour by human activities have only a negligible impact on the concentration in the troposphere.

The heating effect of infrared absorption

When bonds in the molecules of greenhouse gases absorb infrared radiation from the Earth, they vibrate with greater energy.

This absorbed energy can be re-emitted in the form of infrared radiation. However, when this happens, the infrared radiation is emitted in all directions. The overall effect of this is that some of the radiation that would have been lost into space is now returned to Earth, causing heating of the surface.

Some of the vibrational energy is also transferred to other molecules in the atmosphere by collisions. This increases the average kinetic energy of molecules in the atmosphere and the temperature of the atmosphere increases.

Putting it all together

Figure 3 helps visualise the greenhouse effect.

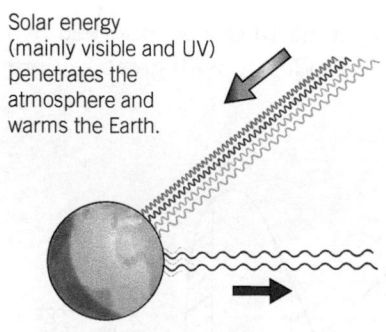

Solar energy (mainly visible and UV) penetrates the atmosphere and warms the Earth.

The Earth absorbs this energy, heats up and radiates IR Greenhouse gases in the troposphere absorb some of this IR

▲ **Figure 3** *The greenhouse effect*

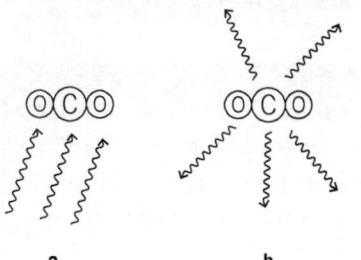

a   b

▲ **Figure 2** **a** *Greenhouse gases absorb infrared radiation emitted from the Earth* **b** *the infrared radiation is re-emitted in all directions*

Model answer: The greenhouse effect

Describe the greenhouse effect and how an increase in carbon dioxide concentration in the troposphere is responsible for an enhanced greenhouse effect.

- The Sun emits mostly visible and ultraviolet radiation.
- Most of the ultraviolet radiation is absorbed by the Earth's atmosphere, but visible light reaches the surface of the Earth.
- This visible light is absorbed by the Earth's surface and warms it up.
- The Earth's surface emits infrared radiation.
- Some frequencies of this infrared radiation are absorbed by carbon dioxide molecules in the troposphere, increasing the vibrational energy of the bonds; the remaining frequencies (in the infrared window) escape into space.
- The absorbed energy is re-emitted as infrared radiation in all directions .
- Some of this is emitted in the direction of the Earth's surface, heating it up.
- Some energy is also transferred (by collisions) to other molecules in the atmosphere, increasing the temperature of the atmosphere.
- If carbon dioxide concentration increases, more infrared radiation is absorbed and less energy is lost into space through the infrared window.

> Always start by describing the ultimate source of all the energy reaching the Earth's atmosphere — the Sun.

> Describe the processes causing the Earth to emit infrared radiation.

> Describe how molecules of greenhouse gas absorb infrared radiation.

> Describe how the absorbed energy causes heating of the atmosphere and the Earth's surface.

> Make a clear link between increased concentrations of greenhouse gases and an increase in energy absorbed by these gases.

Summary questions

1 Describe the differences in the electromagnetic radiation emitted by the Sun and the Earth. In your answer include reference to wavelength and frequency. *(3 marks)*

2 Sulfur hexafluoride (SF_6) is a powerful greenhouse gas because it absorbs in the infrared window. Explain the meaning of this statement. *(2 marks)*

3 If the temperature of the Earth rises, more water will evaporate to form water vapour. Describe and explain the further effect this will have on the temperature of the Earth. *(5 marks)*

Revision tip

When describing the greenhouse effect you should begin by describing the radiation emitted by the Sun.

1 The infrared window is the range of wavelengths of infrared radiation that are:

   A Emitted by the Earth's surface

   B Absorbed by carbon dioxide

   C Absorbed the Earth's surface

   D Not absorbed by water vapour   (*1 mark*)

2 After molecules of a greenhouse gas absorb infrared radiation, the molecules:

   A Collide more frequently

   B Reflect the radiation back to Earth

   C Pass on the energy by colliding with other molecules

   D Have electrons that have been excited to higher energy levels   (*1 mark*)

3 Dinitrogen oxide, N_2O is a powerful greenhouse gas. The best explanation of this is likely to be:

   A It has very polar bonds so absorbs infrared radiation very efficiently.

   B Its concentration is increasing rapidly due to human activities.

   C It absorbs radiation in the infrared window.

   D It is formed from reactions in the soil so is present in high concentrations close to the Earth's surface.   (*1 mark*)

4 Which of the following statements is true about radiation from the Earth and the Sun?

   1 Radiation from the Sun has a longer wavelength than radiation from the Earth.

   2 Radiation from the Sun is mostly in the form of visible and infrared radiation.

   3 The amount of energy radiated by the Earth is similar to the amount of energy that it absorbs from the Sun.

   A 1,2, and 3   C 2 and 3

   B 1 and 2   D Only 3   (*1 mark*)

5 The enhanced greenhouse effect describes:

   A The extra warming of the Earth due to human activities

   B The extra warming of the Earth due to gases other than carbon dioxide

   C The warming of the Earth that has occurred in the last century

   D The warming of the Earth due to carbon dioxide in the atmosphere   (*1 mark*)

6 The temperature of the Earth has increased by about 1°C since 1850. The concentration of carbon dioxide has increased from 270 ppm to 400 ppm. Scientists believe that the increase in carbon dioxide has caused most of this warming. Explain how an increase in carbon dioxide causes warming of the Earth's surface and atmosphere.   (*6 marks*)

16.1 DNA and RNA

Specification reference: PL (c), PL (d)

Nucleic acids

Nucleic acids are condensation polymers formed from monomers called **nucleotides**.

The two types of nucleic acids found in living cells are DNA (deoxyribonucleic acid) and RNA (ribonucleic acid).

Nucleotides

Nucleotide molecules have three components:

- a phosphate group
- a sugar: deoxyribose (in DNA) or ribose (in RNA)
- a 'base': cytosine, adenine, guanine, or thymine (thymine is replaced by uracil in RNA).

Revision tip

You do not need to know the full names of these nucleic acids – the abbreviations are always acceptable as a way of naming them.

Synoptic link

Condensation polymers are described in Topic 13.8, Amino acids, peptides, and proteins.

a Monomers of DNA and RNA

phosphate   ribose   deoxyribose

uracil   cytosine   adenine   guanine

(thymine has a CH₃ at position*)

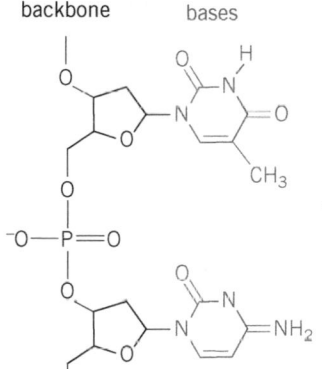

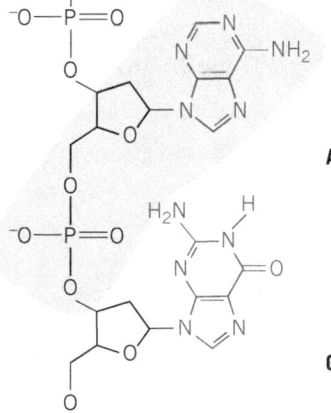

▲ **Figure 1 a** *the components of DNA and RNA nucleotides* **b** *A chain of nucleotides in a DNA strand. The shaded section shows the arrangement of phosphate, sugar, and base in an individual nucleotide*

Formation of nucleotides from subunits

Nucleotides are formed from the three subunits by condensation reactions in which a molecule of water is lost.

Revision tip

You need to be able to identify the atoms in the subunits that are involved in the condensation reactions that form nucleotide molecules.

Joining a sugar to a base

A nitrogen atom from the base bonds to a carbon atom in the sugar molecule. The hydroxyl group from the sugar and a hydrogen atom from the base combine to form a water molecule.

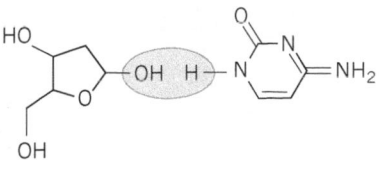

▲ **Figure 2** *A water molecule is lost when a sugar and a base bond together*

Joining a sugar to a phosphate

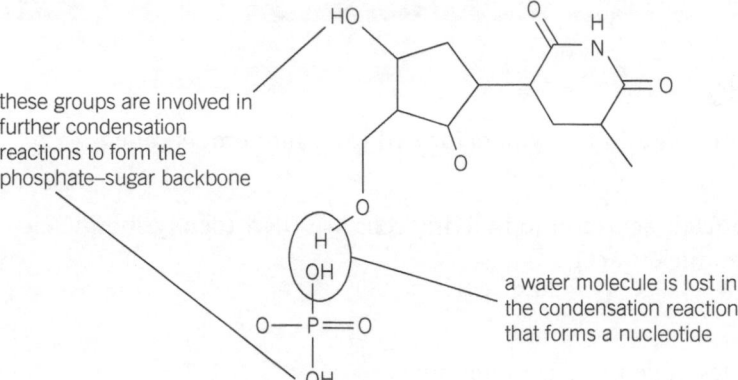

these groups are involved in further condensation reactions to form the phosphate–sugar backbone

a water molecule is lost in the condensation reaction that forms a nucleotide

▲ **Figure 3** *Water molecules are lost when a sugar and phosphate groups bond together*

Key term

Phosphate–sugar backbone: The pattern of alternating phosphate and sugar molecules that joins nucleotides together in nucleic acids.

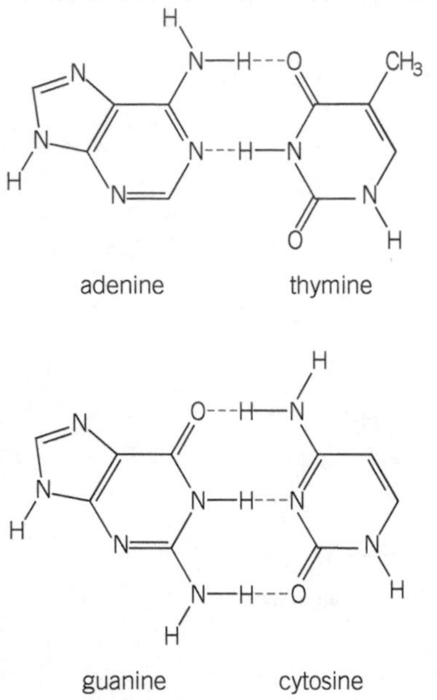

adenine   thymine

guanine   cytosine

▲ **Figure 4** *Complementary base-pairs formed between bases in DNA*

Revision tip

You need to be able to see how the atoms in the complementary bases are arranged in exactly the right places to allow hydrogen bonds to form. Make sure that you can identify the atoms in the two bases that are involved in the hydrogen bonds.

Key term

Codon: A series of three nucleotide bases (e.g. GCC) in a DNA or RNA molecule that codes for an amino acid in a protein.

An oxygen atom from the sugar molecule bonds to the phosphorus atom in the phosphate group to form a nucleotide, The hydrogen atom from the sugar and a hydroxyl group from the phosphate combine to form a water molecule.

The remaining OH groups on the phosphate and sugar then take part in a further condensation reaction to form the phosphate–sugar backbone.

The structure of nucleic acids

Nucleotides bond together to produce nucleic acids by forming a phosphate–sugar backbone.

The base remains attached to the sugar molecule in the phosphate–sugar backbone.

RNA usually consists of a single nucleic acid strand, often twisted into the shape of a helix.

DNA consists of two nucleic acid strands bonded together into a double helix by complementary base pairing.

Complementary base-pairing

The bases in a DNA molecule are in the centre of the double helix structure of the molecule.

Hydrogen bonds form between particular pairs of bases. These hydrogen bonds hold the two strands of DNA together.

Adenine (A) always bonds to thymine (T) by two hydrogen bonds.

Guanine (G) always bonds to cytosine (C) by three hydrogen bonds.

Genetic information

The sequence of bases in a DNA molecule determines the structure of the proteins that are synthesised by cells. Information stored in DNA is described as genetic information. Each codon in the DNA molecule codes for a specific amino acid in the protein.

Copying (replication) of genetic information

A DNA molecule (and the genetic information stored within it) can be copied (a process known as **replication**):

- The DNA molecule 'unzips'; the hydrogen bonds between the complementary base pairs are broken and the two strands separate.
- New DNA nucleotides bond to the exposed bases forming new complementary base pairs.

- The newly attached nucleotides are joined together, creating a new strand of DNA.
- The end result is two identical copies of the original DNA molecule.

How DNA encodes for RNA (transcription)

In order for proteins to be synthesised, a molecule of RNA, called messenger RNA (mRNA), is formed from a sequence of bases on one of the DNA strands:

- The DNA molecule 'unzips'; the hydrogen bonds between the complementary base pairs are broken and the two strands separate.
- New RNA nucleotides bond to the exposed bases forming new complementary base pairs.
- The newly attached nucleotides are joined together, creating a new strand of RNA.
- The RNA molecule detaches from the DNA strand.
- The DNA molecule 'zips' back up.

How RNA encodes for proteins (translation)

The sequence of codons in a messenger RNA molecule determines the sequence of amino acids in a protein. This process involves a second type of RNA molecule known as a transfer RNA (tRNA). One end of the tRNA molecule bonds to a specific amino acid, and the other end has a sequence of three bases called an **anticodon**.

- A mRNA molecule diffuses out of the nucleus.
- It attaches to a binding site in a structure in known as a ribosome.
- A tRNA molecule delivers a specific amino acid to the binding site of the ribosome.
- The anticodon on the tRNA bonds to the complementary codon on the mRNA molecule.
- A second tRNA delivers an amino acid to the second binding site of the ribsome.
- A peptide bond forms between the two adjacent amino acids.
- The tRNA molecules detach from the growing protein chain.

  The ribosome moves along the mRNA molecule and the process repeats.

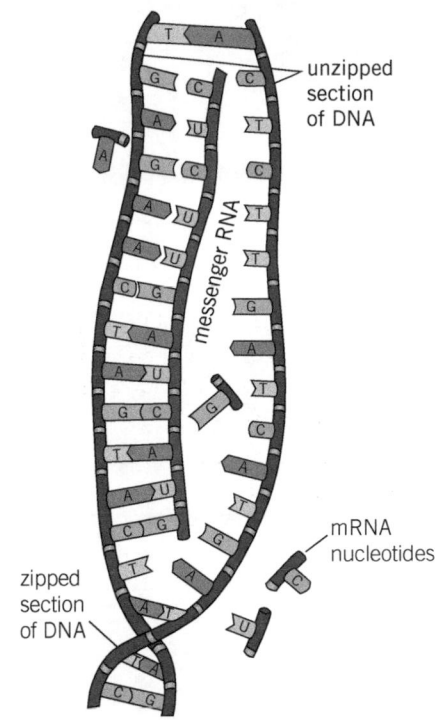

▲ **Figure 5** *During the process of transcription a mRNA molecule forms on one strand of a DNA molecule*

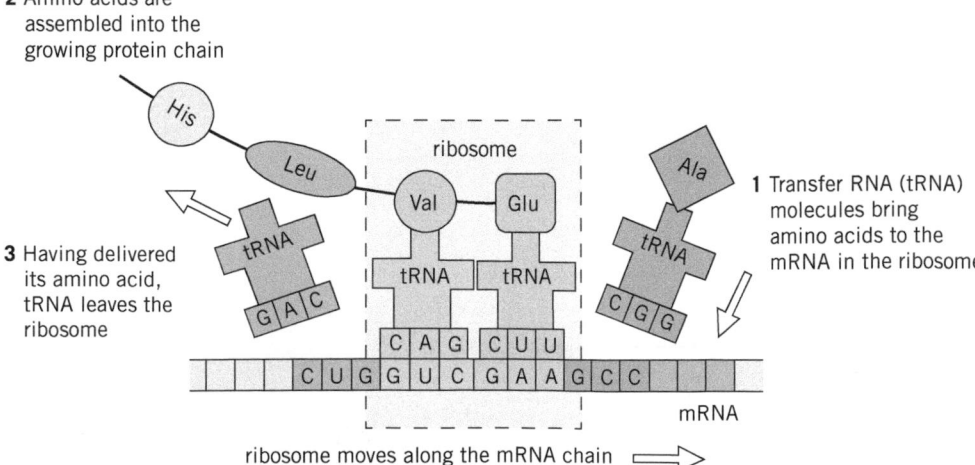

▲ **Figure 6** *The sequence of codons on the mRNA molecule are translated into a sequence of amino acids in a protein chain*

The genetic code

The four bases in RNA (or DNA) can be combined in 64 different ways to make a triplet codon. These codons must code for 20 different amino acids, so one amino acid may be coded for by 3 or even 4 codons.

 Go further

The genetic code is based on triplet codons. There are enough possible combinations of bases to code for all 20 naturally occurring amino acids.

A system based on duplet codons (consisting of 2 bases) could not do this.

Calculate the number of combinations of the four bases in RNA that are possible in a duplet codon system.

Summary questions

1 **a** Describe the structure of DNA. (4 marks)

   b State the three differences between the structure of DNA and RNA. (3 marks)

2 Deduce the amino acid sequence that would be formed in a protein synthesised from the following mRNA base sequence: GCCCUGCAAGUC (1 mark)

3 Discuss the roles that mRNA and tRNA play in the synthesis of proteins. (4 marks)

Chapter 16 Practice questions

1 A molecule of DNA:

 A Contains four bases described by the letters C,G,T,U

 B Consists of a backbone of sugar molecules to which bases are bonded

 C Has hydrogen bonds holding sugar molecules to phosphate groups

 D Has an overall negative charge due to the presence of phosphate groups

 (1 mark)

2 During the process by which genetic information is replicated:

 A Free DNA nucleotides hydrogen bond to bases on a DNA strand

 B Free bases bond to sugar molecules on a DNA strand

 C Covalent bonds between bases break, forming a single strand of DNA

 D A DNA strand with a complementary base sequence forms hydrogen bonds to the bases on a single DNA strand *(1 mark)*

3 In which of these processes do condensation reactions occur?

 1 The formation of the phosphate–sugar backbone.

 2 The attaching of bases to deoxyribose–sugar molecules.

 3 The formation of complementary base pairs. *(1 mark)*

 A 1,2, and 3   C 2 and 3

 B 1 and 2   D Only 3 *(1 mark)*

4 RNA has several structural similarities to DNA. However, differences in the structure of RNA include:

 1 There is only one hydrogen bond formed per base pair in RNA.

 2 RNA has a uracil base in place of a cytosine base.

 3 RNA has a ribose sugar in place of a deoxyribose sugar. *(1 mark)*

 A 1,2, and 3   C 2 and 3

 B 1 and 2   D Only 3

5 Complementary base pairing occurs between pairs of bases, for example between guanine and cytosine.

 The most important reason for this is:

 A The two molecules have a similar shape.

 B Several hydrogen bonds can form between this pair of bases.

 C Ionic bonds can form between an acidic group in one molecule and a basic group in the other.

 D A condensation reaction can occur between these two molecules. *(1 mark)*

6 Part of the amino acid sequence in a protein is:

 Glycine-serine-valine

 The sequence of **DNA** codons that would encode for this sequence is:

 A CCAAGCCAG   C GGTTCGGTC

 B GGUUCGGUC   D CCTTGCCTG

7 DNA and RNA are nucleic acids, found in living cells.

 A student finds this description of the function of DNA and RNA in protein synthesis:

 "DNA encodes for RNA, which codes for an amino acid sequence in a protein"

 a Describe the process by which DNA encodes for RNA. *(3 marks)*

 b The diagram below illustrates how RNA encodes for an amino acid sequence in a protein:

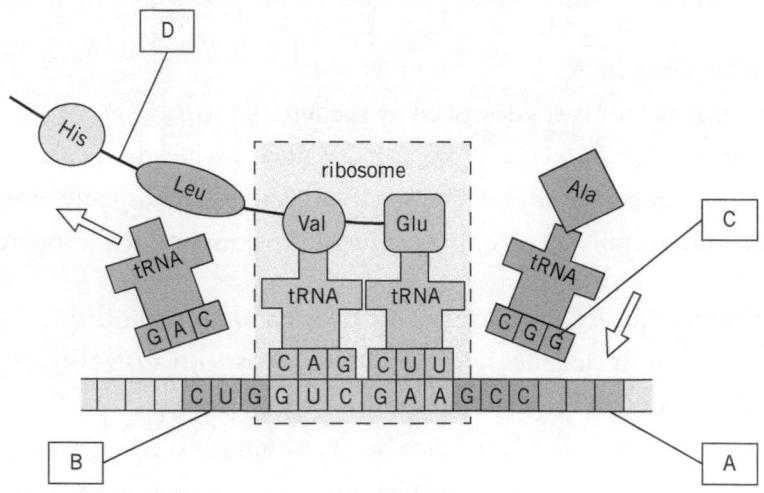

Identify the following structural features:

i   The molecule labelled A. *(1 mark)*

ii The set of three bases CUG labelled B. *(1 mark)*

iii The set of three bases CGG labelled C. *(1 mark)*

iv The functional group, D that forms when amino acids are joined in this process. *(1 mark)*

c Explain how the RNA molecule, which consists of a sequence of just four different bases, is able to encode for a protein which contains 20 different amino acids. *(2 marks)*

17.1 Functional group reactions

Specification reference: CD(f), CD(j)

Polyfunctional molecules

An example of a polyfunctional molecule is an amino acid. It contains two different functional groups, $-NH_2$ and $-COOH$.

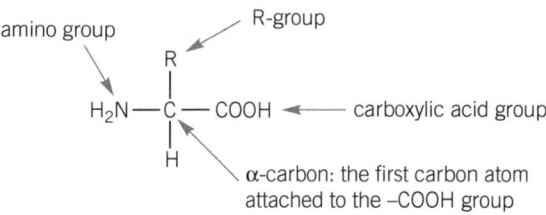

▲ **Figure 1** *An amino acid*

To predict the properties of a polyfunctional molecule, you need to be able to identify the functional groups in the molecule and predict their properties. Generally, you can assume that the properties of the functional groups in a polyfunctional molecule is the same as they are in individual molecules.

Functional groups

This is a list of various functional groups, and their chemical behaviour.

Carbon-based

● Alkenes, C=C: undergo addition reactions. See Topic 12.2, Alkenes.

Halogen-based

● Haloalkanes, R–X: undergo substitution reactions with nucleophiles. See Topic 13.2, Haloalkanes.

Oxygen-based

● Alcohols, R–OH: undergo oxidation to carbonyl compounds and/or carboxylic acids; undergo esterification; undergo nucleophilic substitution reactions with halides; undergo dehydration to form alkenes. See Topic 13.3, Alcohols.

● Carboxylic acids, R–COOH: undergo acid reactions with metals, hydroxides, and carbonates; undergo esterification; undergo conversion to acyl chlorides with $SOCl_2$. See Topic 13.4, Carboxylic acids and phenols, and 13.5, Carboxylic acids.

● Phenols, C_6H_5–OH: undergo acid reactions with hydroxides; undergo esterification with acyl chlorides. See Topic 13.4, Carboxylic acids and phenols.

● Esters, R–COOR: undergo hydrolysis to form alcohols and carboxylic acids. See Topic 13.5, Carboxylic acids, and 13.7, Hydrolysis of esters and amides.

● Acyl chlorides, R–COCl: undergo esterficiation with phenols and alcohols; undergo amide formation with amines. See Topic 13.6, Amines.

● Aldehydes, R–CHO, and ketones, R–CO–R: undergo nucleophilic addition with HCN; undergo oxidation to carboxylic acids (aldehydes only). See Topic 13.10, Aldehydes and ketones.

Nitrogen-based

● Amines, $R-NH_2$: undergo base reactions with acids; act as nucleophiles with acyl chlorides. See Topic 13.6, Amines.

Key term

Polyfunctional molecule: A molecule with more than one functional group.

Synoptic link

You learned about amino acids in Topic 13.8, Amino acids, peptides, and proteins.

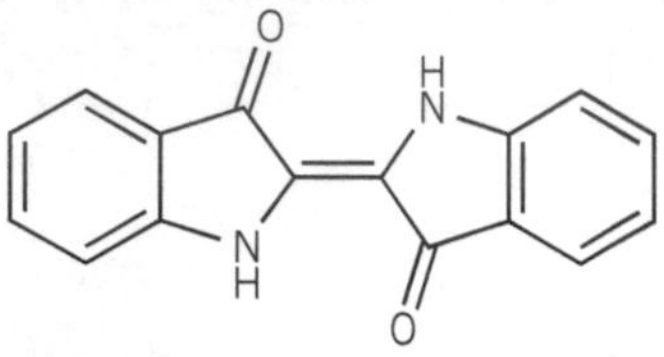

▲ **Figure 2** *Indigo*

- Nitro group, $R–NO_2$: undergoes reduction to amines with tin and concentrated hydrochloric acid.
- Nitriles, $R–CN$: undergo hydrolysis to carboxylic acids.
- Amides, $R–CONH–R$: undergo hydrolysis to carboxylic acids and amines.

Examples of polyfunctional molecules

Indigo (see Figure 2) has the following functional groups:

- **Ketone**, $R–CO–R$, so it could be reduced to a secondary alcohol.
- **Secondary amine**, $R–NH–R$, so it acts as a base by accepting H^+ ions, and it could act as a nucleophile.
- **Alkene**, $C=C$, so it will undergo electrophilic addition reactions.

3-nitrobenzoic acid has the following functional groups:

- **Nitro group**, $R–NO_2$, so it can be reduced to an amine.
- **Carboxylic acid**, $R–COOH$, so it can donate protons and react with metals and bases, form esters by reacting with alcohols, and be converted to an acyl chloride with $SOCl_2$.

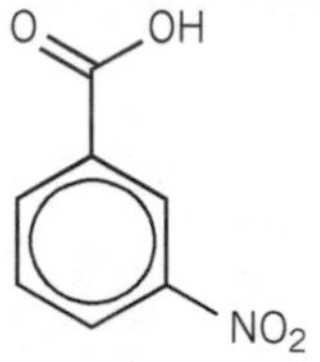

▲ **Figure 3** *3-nitrobenzoic acid*

Synthetic routes

Using your knowledge of the reactions of the functional groups above, it is possible to plan a synthesis of a target molecule. This involves identifying the functional groups in the target and considering how they can be formed from precursors.

Generally a number of intermediates will be produced. For a polyfunctional molecule, it is important to check that the proposed reagents to carry out one conversion do not react with the other functional groups present. If so, a different approach may be required, or the steps may need to be done in a different order.

Starting molecule → intermediate(s) → target molecule

Summary questions

1   Define the term 'polyfunctional molecule'. (*1 mark*)

2   List the functional groups which can undergo:
    a   addition reactions
    b   substitution reactions
    c   oxidation reactions. (*3 marks*)

3   Draw a molecule of 3-aminobenzoic acid, identify its functional groups, and suggest the properties of the molecule. (*4 marks*)

17.2 Classification of organic reactions

Specification reference: CD(j), CD(l)

Organic mechanisms

You have learnt about a wide range of organic reactions during this course. They are summarised below.

Addition reactions

In an addition reaction, two molecules react to form a single product. The atom economy is 100% as there is no waste product. Examples include electrophilic addition of bromine to an alkene (Reaction 1), or nucleophilic addition of HCN to a carbonyl compound (Reaction 2):

Reaction 1: $C_2H_4 + Br_2 \rightarrow C_2H_4Br_2$ see Topic 12.2

Reaction 2: $CH_3CHO + HCN \rightarrow CH_3CH(OH)CN$ see Topic 13.10

Condensation reactions

Two molecules react to form a larger molecule, and a small molecule, such as H_2O or HCl, is removed. The atom economy is less than 100%. Examples include the formation of esters (Reaction 3), and the formation of amides (Reaction 4):

Reaction 3: $CH_3COOH + C_2H_5OH \rightarrow CH_3COOC_2H_5 + H_2O$ see Topic 13.3

Reaction 4: $C_6H_5COCl + CH_3NH_2 \rightarrow C_6H_5CONHCH_3 + HCl$ see Topic 13.6

Elimination reactions

A small molecule, such as HCl or H_2O, is removed from a larger molecule, leaving an unsaturated molecule. The atom economy is less than 100%. Examples include elimination of hydrogen halides from haloalkanes (Reaction 5) or dehydration of alcohols (Reaction 6):

Reaction 5: $C_2H_5Br \rightarrow C_2H_4 + HBr$ see Topic 13.2

Reaction 6: $C_2H_5OH \rightarrow C_2H_4 + H_2O$ see Topic 13.3

Substitution reactions

A group of atoms takes the place of another group in a molecule. The atom economy is less than 100% unless both products are useful. Examples include nucleophilic substitution of haloalkanes (Reaction 7), the reaction of alcohols with halides (Reaction 8), and electrophilic substitution of arenes (Reaction 9):

Reaction 7: $C_3H_7Br + OH^- \rightarrow C_3H_7OH + Br^-$ see Topic 13.2

Reaction 8: $C_4H_9OH + HBr \rightarrow C_4H_9Br + H_2O$ see Topic 13.3

Reaction 9: $C_6H_6 + C_2H_5Cl \rightarrow C_6H_5(C_2H_5) + HCl$ see Topic 12.4

Oxidation reactions

Oxygen atoms are gained and/or hydrogen atoms are lost. Examples include the oxidation of primary alcohols to aldehydes (Reaction 10), and the oxidation of aldehydes to carboxylic acids (Reaction 11):

Reaction 10: $C_3H_7OH + [O] \rightarrow C_2H_5CHO + H_2O$ see Topic 13.3

Reaction 11: $C_2H_5CHO + [O] \rightarrow C_2H_5COOH$ see Topic 13.10

Reduction reactions

Oxygen atoms are lost and/or hydrogen atoms are gained. An example is the reduction of carbonyl compounds to alcohols (Reaction 12):

Reaction 12: $CH_3COCH_3 + 2[H] \rightarrow CH_3CH(OH)CH_3$

> **Synoptic link**
>
> You covered atom economy in Topic 14.2, Atom economy.

> **Key term**
>
> **Electrophile:** A species that accepts a pair of electrons to form a covalent bond. Examples include Br^+ or $Br^{\delta+}$.

> **Revision tip**
>
> C_2H_4 is ethene, an alkene, so it has a C=C bond, and is unsaturated.

> **Key term**
>
> A nucleophile is a species with a lone pair that can form a covalent bond. Examples include Br^- and H_2O.

> **Revision tip**
>
> $C_6H_5(C_2H_5)$ is ethylbenzene.

> **Revision tip**
>
> Remember that [O] represents an oxygen from the oxidising agent.

> **Revision tip**
>
> [H] represents a hydrogen from the reducing agent.

Hydrolysis reactions

Bonds are broken by reaction with water. Examples include the hydrolysis of esters (Reaction 13), and the hydrolysis of amides (Reaction 14), which can be catalysed by acids or alkalis:

Reaction 13: $C_2H_5COOCH_3 + H_2O \rightarrow C_2H_5COOH + CH_3OH$ see Topic 13.8

Reaction 14: $CH_3CONHCH_3 + H_2O \rightarrow CH_3COOH + CH_3NH_2$ see Topic 13.8

A summary of organic reactions

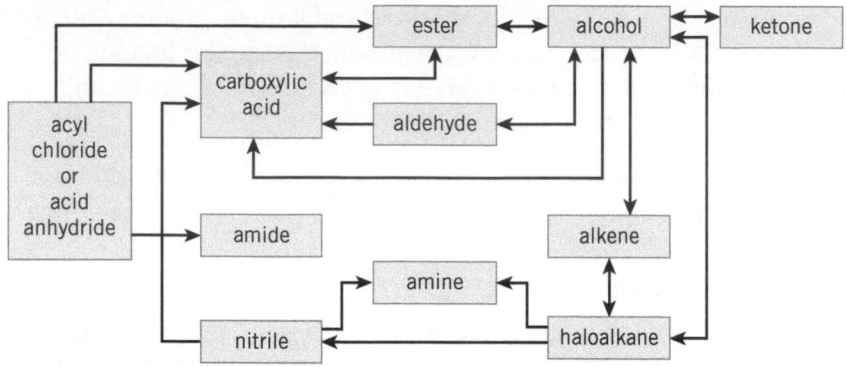

▲ **Figure 6** *Some important functional group interconversions*

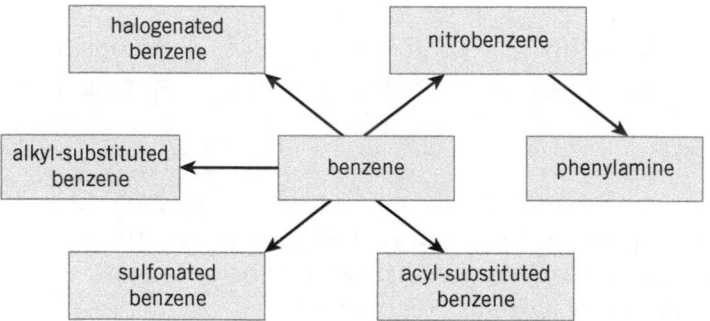

▲ **Figure 7** *Functional group interconversions for aromatic compounds*

Synthetic routes

Using the charts above it is possible to plan potential synthesis reactions.

 Worked example: Planning a synthetic route

Suggest a synthetic route to produce ethylamine from ethanol.

Step 1: Look at the chart to find a route between an alcohol and an amine:

alcohol → alkene → haloalkane → amine

Step 2: Identify the specific compounds involved:

ethanol → ethene → bromoethane → ethylamine

Step 3: Consider the reagents and conditions needed:

alcohol → alkene: dehydration with Al_2O_3 at 300 °C

alkene → haloalkane: reaction with HBr

haloalkane → amine: reaction with NH_3

Chapter 17 Practice questions

1 Which of the following is a polyfunctional molecule?

   A 2-methylpentane

   B pentanedioic acid

   C pentyl ethanoate

   D glycine *(1 mark)*

2 Which of the following undergoes nucleophilic addition reactions?

   A carbonyl compounds

   B arenes

   C haloalkanes

   D alkenes *(1 mark)*

3 Which of the following reacts with NaOH?

   A phenol

   B ethanol

   C propene

   D nitrobenzene *(1 mark)*

4 Which of the following can react to form a carboxylic acid in one step?

   1 nitriles

   2 ketones

   3 alkenes

      A 1 only

      B 1 and 2

      C 2 and 3

      D 1, 2, and 3 *(1 mark)*

5 Which of the following reacts with HBr in an electrophilic addition reaction?

   A propene

   B propan-1-ol

   C propylbenzene

   D propanone *(1 mark)*

6 Identify the functional groups in Disperse Red 60 dye, and suggest the chemical properties of the molecule. *(7 marks)*

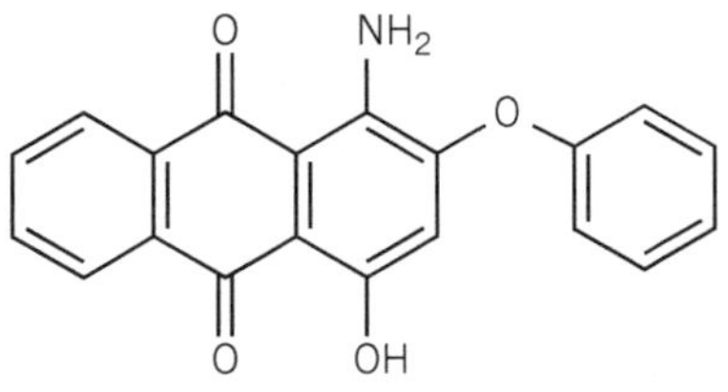

▲ **Figure 8** *Disperse Red 60 dye*

7 Suggest a synthetic route to make $CH_3CH_2CH_2CN$ from CH_3CH_2CHO, giving the intermediates, and reagents and conditions. *(3 marks)*

8 Suggest a synthetic route to make phenylamine from benzene in two steps, giving the intermediate, and reagents and conditions. *(2 marks)*

Answers to summary questions

4.4

1 enthalpy

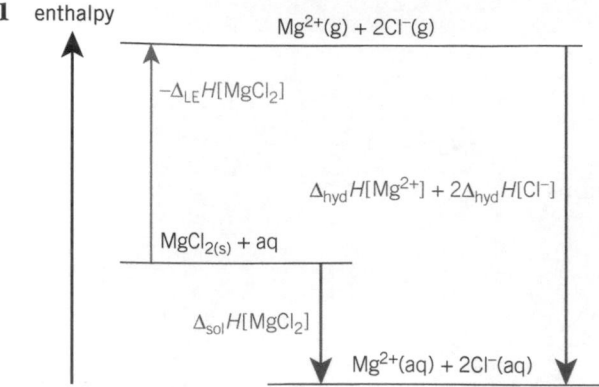

Correct substances on lines *[1 for each]*; correct positions of lines *[1 for each]*; correct labels on each arrow. *[1 for each][Total 7]*

2 For calcium chloride to dissolve, ionic bonds between calcium ions and chloride ions must break *[1]*; only weak ion–dipole bonds can form *[1]*; bonds broken would be stronger than bonds formed *[1]*; energy released by forming bonds would not compensate for energy needed to break bonds. *[1]*

3 a Sum of enthalpy changes of hydration = enthalpy change of solution + lattice enthalpy *[1]*

$= -155 - 2526 = -2681 \, \mathrm{kJ\,mol^{-1}}$ *[1]*

So $\Delta_{hyd}H \, [Mg^{2+}(aq)] + 2 \Delta_{hyd}H \, [Cl^-(aq)] = -2681$ *[1]*

$2 \Delta_{hyd}H \, [Cl^-(aq)] = -2681 - (-1926) = 755$;

$\Delta_{hyd}H \, [Cl^-(aq)] = -\dfrac{755}{2} = -377.5 \, \mathrm{kJ\,mol^{-1}}$ *[1]*

b Ca^{2+} ions have a smaller charge density (than Mg^{2+}) *[1]* weaker ion-dipole bonds are formed between Ca^{2+} and water molecules than between Mg^{2+} and water molecules. *[1]*

4.5

1 Negative *[1]*; there are fewer moles of gas on the RHS of the equation *[1]*; so there are fewer ways of arranging the particles after the reaction *[1]*; this means that entropy decreases (so ΔS is negative). *[1]*

2 $\Delta_{sys}S = \Sigma S \text{ (products)} - \Sigma S \text{ (reactants)}$ *[1]* $= (2 \times 26.9) + (4 \times 240) + 205 - (2 \times 164) = +890.8 \, \mathrm{J\,K^{-1}\,mol^{-1}}$ *[1]*

$\Delta_{surr}S = -\Delta H/T$ *[1]* $= -510.8 \times \dfrac{1000}{298}$
$= -1714 \, \mathrm{J\,K^{-1}\,mol^{-1}}$ *[1]*

$\Delta_{tot}S = \Delta_{sys}S + \Delta_{surr}S = 890.8 - 1714$
$= -823.2 \, \mathrm{J\,K^{-1}\,mol^{-1}}$ *[1]*; reaction is not feasible / spontaneous at this temperature. *[1]*

3 At room temperature $\Delta_{tot}S$ is negative (because the reaction is not feasible) *[1]*; reaction is feasible at certain temperatures so $\Delta_{surr}S$ and $\Delta_{sys}S$ mist have opposite signs *[1]*; reaction becomes feasible when temperature is increased, so $\Delta_{surr}S$ must be negative *[1]*; $\Delta_{sys}S$ must be positive. *[1]*

5.5

1 Secondary structure is the folding of the chain into 3-dimensional structures (helix and sheet) *[1]*; tertiary structure is the further folding into the overall complex 3-dimensional structure of the enzyme *[1]* secondary structure is maintained by hydrogen bonds between peptide groups *[1]* tertiary structure is maintained by various types of bond between the amino acid side groups. *[1]*

2 O of C=O group shown bonding to H atom of NH group (using dotted line to denote hydrogen bond) *[1]* O shown as δ-, H as δ- *[1]* lone pair on O shown pointing in direction of hydrogen bond. *[1]*

3 Changes in primary structure means that different amino acids may be present in active site *[1]* this affects the pattern of bonding that maintains the shape of the active site *[1]* different shaped active site may no longer be complementary to the substrate of the enzyme OR different pattern of amino acids may affect ability of active site to bind to substrate. *[1]*

5.6

1 a the part of a molecule that is responsible for its pharmacological / biological activity. *[1]*

b to increase effectiveness *[1]*; to reduce side effects *[1]*; OR give specific beneficial properties that might be improved e.g. altered solubility (in fats / blood) *[1]*; altered rate of breakdown in organism. *[1]*

2 Circle round either structure as shown below: *[1]*

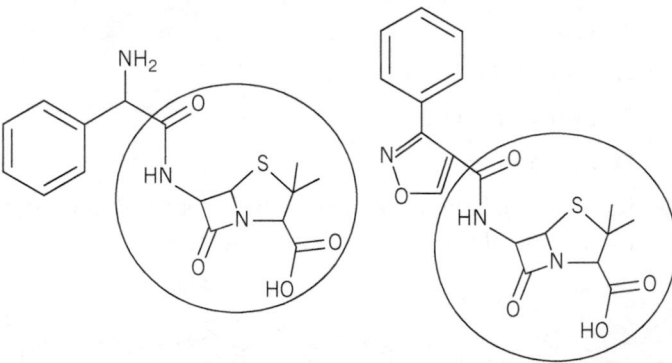

3 a The molecule contains a chiral C atom *[1]*; this is the left-hand carbon atom in the right-hand ring. *[1]*

b The three-dimensional arrangement of the groups in the two enantiomers is different *[1]*; may have a different bonding pattern with a receptor site / active site. *[1]*

5.7

1 hydrogen bonds, instantaneous dipole-induced dipole, ionic, covalent *[1 mark each]*

2 The sulfonic acid group on the acid blue molecule can ionise to form SO_3^- *[1]*; NH_2 groups gain H^+ ions to form NH_3^+ *[1]*; ionic bonds form between SO_3^- and NH_3^+ *[1]*

3 Hydrogen bonds form between the dye and cotton [1]; because of the presence of OH groups in both molecules [1]; there are hydrogen bonds between water molecules [1]; a few / weak hydrogen bonds form between the dye and water [1]; the bonds that can be formed are weaker than the bond that need to break [1]; the energy released from bond formation is not sufficient to compensate for bond breaking. [1]

6.6

1 High-resolution mass spectrometry measures m/z (or relative mass) of M+ peak to 4 d.p. [1]; individual atoms / isotopes have relative masses that are not whole numbers [1]; so each molecular formula has an M+ peak with a unique mass [1]; identify molecule by comparing with database. [1]

2 a molecular mass = 72, so 57 is formed by loss of 15 = CH_3 [1]; 43 is formed by loss of 29 = C_2H_5 [1]; 29 is formed by loss of 43 = CH_3CO [1]

b Formulae of fragments: 57 = $C_3H_5O^+$ or $CH_3CH_2CO^+$ [1]; 43 = $C_2H_3O^+$ or $COCH_3^+$ [1]; 29 = $C_2H_5^+$ or $CH_3CH_2^+$ [1]

3 M$^+$ peak will be same (88) for both isomers in low resolution [1] and in high resolution [1]; some fragments will be identical because same groups are present, e.g. at 15 (CH_3^+) and 29 ($CH_3CH_2^+$) [1]; some fragments will be identical because different groups may have same mass, e.g. 43 ($C_3H_7^+$ in propyl ethanoate and CH_3CO^+ in ethyl ethanoate) , 45 ($HCOO^+$ for propyl methanoate, $OC_2H_5^+$ for ethyl ethanoate) [1]; there are no fragments that are likely to be unique to one of these isomers [1]; however, the heights of the fragment peaks may be different database could be used to identify molecule). [1]

6.7

Go further

Three protons can be aligned in four different ways:

all protons with field ↑↑↑, all protons against field ↓↓↓, 2 protons with field + 1 proton against field ↑↑↓, 2 protons against field + 1 proton with field ↓↓↑

The ratios of heights of the multiplet peaks will be 1 : 3 : 3 : 1

Summary questions

1 a 2 peaks [1]   **b** 3 peaks [1]

2 Quartet is produced by a CH_3 group adjacent to the CH proton environment OR there are three H atoms on the C atom that is adjacent to the CH group [1]; doublet is produced by the CH group adjacent to the CH_3 proton environment OR there is one H atom on the C atom that is adjacent to the CH_3 group. [1]

3 a There are 4 proton environments [1]; doublet at δ = 1.2 ppm suggests that there is a CH group adjacent to the CH_3 environment producing this peak [1]; quartet suggests there is a CH_3 group adjacent to the CH environment producing this

peak [1]; peak at δ = 11.0 suggests COO$\underline{H}$ [1]; peak at δ = 2.8 is not split so could be an OH group. [1] [any 4 points] Structure is:

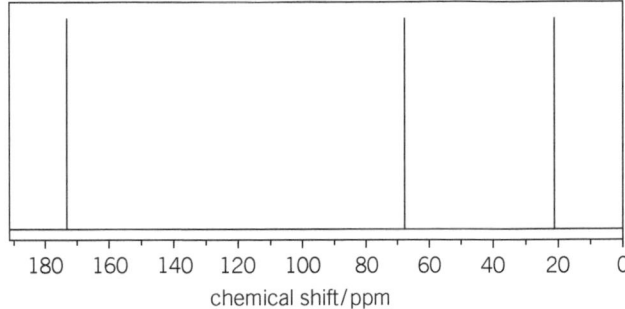

b Three peaks [1]; δ = 180–220 (for $\underline{C}$OOH) [1]; δ = 0–45 (for $\underline{C}H_3$–C) [1]; δ = 50–90 (for $\underline{C}H$–O). [1]

6.8

1 Empirical formula can be deduced from percentage by mass data [1]; relative molecular mass can be deduced from m/z value of M$^+$ peak [1]; combining both data allows molecular formula to be deduced. [1]

2 a mass spectrometry [1]; fragmentation pattern will be different [1]; e.g. fragment at 43 ($C_3H_7^+$) for (i), or 29 ($C_2H_5^+$) for (ii) [1]

b infrared spectrometry [1]; different peaks present [1]; e.g. C=C peak at 1620–1680 cm^{-1} in **i**, OR C=O peak at 1720–1740 cm^{-1} in **ii**, OR O–H peak at 3200–3600 cm^{-1} in **i**. [1]

3 **Points that could be made include:**

Mass spectrum: empirical formula is C_5H_6O; molecular mass (from M$^+$ peak) is 164, hence molecular formula is $C_{10}H_{12}O_2$; Low ratio of H:C suggests an aromatic compound;

IR spectrum: peak at above 1700 suggests C=O in aldehyde, ketone or carboxylic acid; No broad peak at 3200–3600 cm^{-1}, so carboxylic acid not present; Peak at 1620 cm^{-1} could suggest alkene or arene C=C; ^{13}C nmr spectrum: 8 peaks so 8 types of C atom; Low-res ^{1}H nmr spectrum: 5 peaks so 5 different ^{1}H environments Ratio of H atoms in each environment = 2 : 2 : 3 : 2 : 3; 4H peak between δ =7.0 and 8.0 could suggest 4H atoms in arene; Grouping of 2:2 in arene group suggests H atoms on C atoms 2, 3, 5, and 6; High res. ^{1}H spectrum: triplet at δ = 1.2 suggests CH_2 adjacent to CH_3; Quartet at δ = 2.6 suggests CH_3 adjacent to CH_2; Singlet at 3.8 suggests no H atoms on adjacent carbons; could be $\underline{C}H_3O$ Most likely structure is:

Level of response mark scheme:

Marks 5–6: Identifies molecule correctly, linking most features of the molecule clearly to the evidence, AND uses data from all sources of information (MS, IR, ^{13}C nmr, low-res ^1H nmr, and high res ^1H nmr) in reaching conclusion.

Marks 3–4: Identifies some features of the molecule correctly, linking some of the features clearly to the evidence, AND uses data from at least three sources of information in reaching conclusion.

Marks 1–2: Makes limited progress in identifying molecule, with some attempt to link with the evidence, AND uses only one or two sources of information in reaching conclusion.

6.9

1 a the part of the molecule responsible for absorbing visible light (or uv radiation). [1]

 b an extensive conjugated system / system of alternating double and single bonds. [1]

2

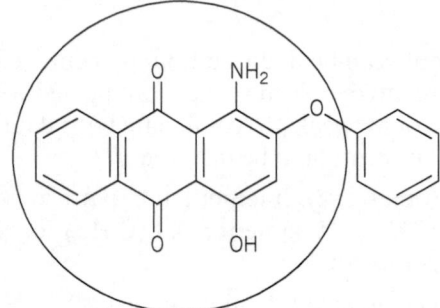

[1 mark for circling correct part of structure]

3 The conjugated system is more extensive in A than B [1]; the energy gap between energy levels in A is smaller than in B [1]; $\Delta E = h\nu$ so A absorbs at a lower frequency than B [1]; A absorbs in the visible region / in the blue part of the spectrum [1]; B absorbs in the uv part of the spectrum. [1]

7.5

1 a $K_C = \dfrac{[COCl_2]}{[CO][Cl_2]}$ [1]

 b i no effect ii K_c decreases iii no effect [3]

2 Increasing the temperature moves the position of equilibrium in the endothermic direction [1], which is to the right [1]; therefore K_c increases. [1]

3 $K_C = \dfrac{[HI]^2}{[H_2][I_2]}$ [1] $= \dfrac{(0.027\,mol\,dm^{-3})^2}{(0.004\,mol\,dm^{-3}) \times (0.004\,mol\,dm^{-3})}$ [1] $= 46$ [1] [no units – 1 mark]

7.6

1 a K_{sp} (AgBr) = $[Ag^+][Br^-]$ [1]

 b K_{sp} (Ag$_2$S) = $[Ag^+]^2[S^{2-}]$ [1]

 c K_{sp} (Ag$_2$CrO$_4$) = $[Ag+]^2[CrO_4^{2-}]$ [1]

2 K_{sp} (AgBr) = $[Ag^+][Br^-]$ [1] = $(7.07 \times 10^{-7}\,mol\,dm^{-3})^2$ [1] = $5.00 \times 10^{-13}\,mol^2\,dm^{-6}$ [1]

3 $[Pb^{2+}][SO_4^{2-}] = (1.45 \times 10^{-4}\,mol\,dm^{-3})^2 = 2.1 \times 10^{-8}\,mol^2\,dm^{-6}$ [1]; this is greater than K_{sp}, so a precipitate will form. [1]

7.7

1 The mass spectrometer will confirm the relative molecular mass of the component. [1]

2 Retention time would decrease. [1]

3 Retention time on the x-axis and three peaks labelled A, B, and C at 3 min, 5.5 min and 6.4 min respectively. [1]

8.2

1 H_2SO_4 and HSO_4^- [1]; OH^- and H_2O [1]

2 pH = 1.00 [1]

3 $[H^+] = K_w / [OH^-] = 2 \times 10^{-13}\,mol\,dm^{-3}$ [1]; pH = 12.7 [1]

4 a $K_a = \dfrac{[H^+][HCOO^-]}{[HCOOH]}$ [1]

 b $H^+ = \sqrt{1.6\times10^{-4}\times0.001} = 4 \times 10^{-4}\,mol\,dm^{-3}$ [1]; pH = 3.4 [1]

8.3

1 Since $\dfrac{[acid]}{[salt]} = 1$, $[H^+] = K_a = 6.3 \times 10^{-5}\,mol\,dm^{-3}$ [1]; pH = 4.2 [1]

2 $[H^+] = 3.4 \times 10^{-6}\,mol\,dm^{-3}$ [1]; pH = 5.47 [1]

3 $[H^+] = 1.8 \times 10^{-4} \times \dfrac{0.01}{0.006} = 3.0 \times 10^{-4}\,mol\,dm^{-3}$ [1]; pH = 3.52 [1]

9.3

1 Cu strip immersed in a solution of Cu^{2+} ions [1]; Ag strip immersed in a solution of Ag^+ ions [1]; both solutions 1.0 mol dm^{-3} [1]; voltmeter [1]; salt bridge. [1]

2 MnO_4^- (aq) + $8H^+$ (aq) + $5Fe^{2+}$ (aq) → Mn^{2+} (aq) + $4H_2O$ (aq) + $5Fe^{3+}$ (aq) *[1 mark for correct species on left- and right-hand sides; 1 mark for balancing]*

3 $E^{\ominus}_{cell} = 0.74\,V$ [1]; the reverse reaction is infeasible as $E^{\ominus}_{cell}$ would be negative. [1]

9.4

1 The oxygen half-reaction involves O_2 and H_2O. [1]

2 Iron goes from 0 to +2, so it is oxidised [1]; oxygen in O_2 goes from 0 to –2, so it is reduced. [1]

3 $E^{\ominus}_{cell} = 1.16\,V$ [1]; this is a greater $E^{\ominus}_{cell}$ than the reaction with iron, so zinc corrodes first. [1]

10.5

1 Rate of reaction is the change of concentration of a reactant or product over a given time. The unit is mol dm^{-3} s^{-1}. [1]

2 Measure the rate of production of hydrogen gas [1], the change in mass as hydrogen is released [1], or the rate of change of pH. [1]

3 If it is not possible to measure a property that continuously changes (such as colour or pH), a sample must be removed and the reaction halted by quenching. *[1]*

10.6

1 A is 2nd order; B is 1st order; the overall order is 3. *[3]*

2 If the line is horizontal, it is zero order *[1]*; if there is a straight line proportional to concentration, it is first order *[1]*; if there is a straight line proportional to concentration², it is second order. *[1]*

3 k = rate / $[X]^2$ *[1]*

$k = 4.2 \times 10^{-3}\,mol\,dm^{-3}\,s^{-1}$ / $(0.01\,mol\,dm^{-3})^2 = 42$ *[1]*

Units are $mol^{-1}\,dm^3\,s^{-1}$ *[1]*

10.7

1 Time on *x*-axis and concentration on *y*-axis *[1]*; downwards curve *[1]*; two half-lives shown – one at 50% concentration and one at 25% concentration. *[1]*

2 Rate = k [A] [H⁺]. *[1 mark for each correct order.]* A and H⁺ first order; B zero order.

3 Yes, because A and H appear in the rate equation and in the rate determining step. *[1]*

10.8

1 A substrate reversibly fits into the active site and reacts with the enzyme *[1]*; an inhibitor fits into the active site and blocks the substrate from entering. *[1]*

2 The active site changes shape *[1]*; at high temperatures the hydrogen-bonding of the tertiary structure is affected and at high/low pH the ionic interactions of the tertiary structure are affected. *[1]*

3 When the substrate concentration is high, the active sites are fully occupied so an increase in concentration no longer affects the rate. *[1]*

11.4

1 The total charge is $(2+) + (4 \times 0) = 2+$. *[1]*

2 Addition of ammonia causes a blue precipitate *[1]* of $Cu(OH)_2$ *[1]*. Excess ammonia gives a dark blue solution *[1]*; of $[Cu(NH_3)_4(H_2O)_2]^{2+}$. *[1]*

3 The ligands cause d-orbitals splitting *[1]*; electrons can be excited to a higher energy level *[1]*; visible light is absorbed of frequency given by $\Delta E = h\nu$ *[1]*; the complementary colour (orange) is transmitted. *[1]*

11.5

Go further

Colour will become darker; forward reaction is exothermic so increasing temperature shifts equilibrium to LHS (endothermic direction); more NO_2 is formed, which is darker in colour than N_2O_4.

Summary questions

1 a dinitrogen oxide / nitrogen(I) oxide

  b nitrate(III)   c nitrogen dioxide / nitrogen(IV) oxide *[1 mark each]*

2 To confirm for ammonium ions: dissolve the solid in water and add NaOH *[1]*; heat gently AND moist red litmus paper (above the level of the liquid) will turn blue *[1]*; to confirm for nitrate (V): dissolve the salt in water and add a spatula measure of Devarda's alloy *[1]*; heat gently AND moist red litmus paper (above the level of the liquid) will turn blue. *[1]*

3 $NH_4^+ + 3H_2O$ *[1]* $\rightarrow NO_3^- + 10H^+$ *[1]* $+ 8e^-$ *[1]*

12.4

Go further

(a) aromatic (10 delocalised e⁻), (b) not aromatic (8 delocalised e⁻), (c) aromatic (6 delocalised e⁻)

Summary questions

1 Between each pair of C atoms there is one σ-bond *[1]*; each C atom has an electron in a p-orbital *[1]*; these 6 electrons are delocalised *[1]*; p-orbital overlap *[1]*; form rings of electron density above and below the plane of the C atoms. *[1]*

2 a i 1-bromo-2,5-dichlorobenzene

  ii 4-methylphenol *[1 mark each]*

b

(i)   (ii)   *[1 mark each]*

3 Benzene has a hexagonal structure with bond angles of 120° *[1]*; this is consistent with the Kekulé model as there are three areas of electron density around each C *[1]*; the hexagon is regular with all C–C bond lengths equal *[1]*; this is not consistent with the Kekulé model as the C=C bonds would be shorter than the C–C bonds. *[1]*

12.5

Go further

a $AlCl_3 + Cl_2 \rightarrow AlCl_4^- + Cl^+$

b $H^+ + AlCl_4^- \rightarrow HCl + AlCl_3$

Summary questions

1 Benzene contains a delocalised system of electrons *[1]*; addition would destroy the delocalised system *[1]* but substitution leaves it intact *[1]*; so the product of substitution is more stable than the product of addition. *[1]*

2 a Cl_2, $AlCl_3$, reflux *[1 mark for each]*

  b conc. nitric acid, conc. sulfuric acid, temperature below 55°C *[1 mark for each]*

3 a Aluminium chloride *[1]*; acts as a catalyst *[1]*; generates a reactive electrophile / CH_3CO^+. *[1]*

b

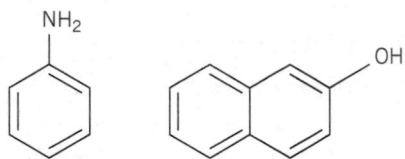

[1 mark for reactants, 2 marks for correct products. Benzene ring can be shown with circle or using Kekule structure.]

12.6

1 Dissolve phenylamine in (ice-cold) hydrochloric acid *[1]*; make up a solution of sodium nitrate (III) in (ice-cold) hydrochloric acid *[1]*; add the two solutions together, ensuring the temperature does not rise above 5°C. *[1]*

2 a to increase solubility OR to allow dye to bond ionically to fibre *[1]*

b to allow dye to bond (via hydrogen bonding) to fibre *[1]*

c to modify the chromophore and hence alter the colour *[1]*

3 Structures are:

[1 mark each, Phenylamine can be shown with a circle or using Kekule structure above.]

13.5

1 a butanoic acid *[1]*   **b** butanedioic acid *[1]*

c benzoic acid *[1]*

2 a $CH_3COOH + K \rightarrow CH_3COO^-K^+ + \frac{1}{2}H_2$ *[1] allow without charges*

b $CH_3COOH + KOH \rightarrow CH_3COO^-K^+ + H_2O$ *[1] allow without charges*

c $2CH_3COOH + K_2CO_3 \rightarrow 2CH_3COO^-K^+ + H_2O + CO_2$ *[1] allow without charges*

3 X is a metal as it produces hydrogen *[1]*. Y is a metal carbonate as it produces carbon dioxide. *[1]*

13.6

1 a $CH_3CH_2CH_2CH_2NH_2$ *[1]*

b $CH_3CH(NH_2)CH_2CH_2CH_3$ *[1]*

c $H_2NCH_2CH_2CH_2NH_2$ *[1]*

2 $2C_2H_5NH_2 + H_2SO_4 \rightarrow 2C_2H_5NH_3^+ + SO_4^{2-}$ *[1 mark for $C_2H_5NH_3^+$; 2 marks if completely correct.]*

3 a $C_6H_5COCl + NH_3 \rightarrow C_6H_5CONH_2 + HCl$ *[1]*

b $C_6H_5COCl + CH_3NH_2 \rightarrow C_6H_5CONHCH_3 + HCl$ *[1]*

c $C_6H_5COCl + NH_3 \rightarrow C_6H_5COOCH_3 + HCl$ *[1]*

13.7

1 $CH_3COOC_2H_5 + H_2O \rightarrow CH_3COOH + C_2H_5OH$ *[1]*

2 $CH_3COOC_3H_7 + OH^- \rightarrow CH_3COO^- + C_3H_7OH$ *[1]*

3 Acid conditions: $CH_3CH_2CONHCH_2CH_3 + H^+ + H_2O \rightarrow$
$CH_3CH_2COOH + CH_3CH_2NH_3^+$ *[1]*

Alkaline conditions: $CH_3CH_2CONHCH_2CH_3 + OH^- \rightarrow$
$CH_3CH_2COO^- + CH_3CH_2NH_2$ *[1]*

13.8

1 a $H_2NCH(CH(CH_3)_2)COO^-$ *[1]*

b $H_3N^+CH(CH(CH_3)_2)COO^-$ *[1]*

c $H_3N^+CH(CH(CH_3)_2)COOH$ *[1]*

2 $H_2NCH(CH_2OH)CONHCH(CH_3)COOH$ *[1]* and
$H_2NCH(CH_3)CONHCH(CH_2OH)COOH$ *[1]*

3 a Acidic conditions give $H_3N^+CH_2COOH \, Cl^- +$
$H_3N^+CH(CH_3)COOH \, Cl^-$; a chloride salt is produced because HCl is the protonating agent (the amine group becomes protonated after the hydrolysis). *[1]*

b Alkaline conditions give a sodium salt of the carboxylic acid group:

$H_2NCH_2COO^-Na^+ + H_2NCH(CH_3)COO^-Na^+$ *[1]*

13.9

1 An ester of propane-1,2,3-triol containing different R groups from three different fatty acids. *[1]*

2

[1]

3 $7.2\,dm^3$ of hydrogen is $0.3\,mol$ *[1]*. Therefore the triester reacts with hydrogen in a 1:2 ratio *[1]*, so there are two C=C bonds. *[1]*

13.10

1 Warm the substances separately with Fehling's solution or with Tollens' reagent. The aldehyde will give a red precipitate with Fehling's solution, or a silver mirror with Tollens' reagent *[1]*. The ketone will not react with either. *[1]*

2 a $CH_3CH_2CHO + [O] \rightarrow CH_3CH_2COOH$. *[1]*

b propanone cannot be oxidised. *[1]*

3

The R group is CH_3. Your diagram must show curly arrows and dipoles *[1]*, the intermediate *[1]* and the final product $CH_3CH(OH)CN$. *[1]*

14.3

1 HCl is a co-product and but-1-ene is a by-product *[1]*; HCl is produced in the intended reaction *[1]*; but-1-ene is produced in a second, unintended reaction. *[1]*

2 A has a higher atom economy *[1]*; so less waste to dispose of *[1]*; A is carried out at a lower temperature *[1]*; so less energy needed for fuel *[1]*; starting material for A is ethene *[1]*; which can be obtained easily from crude oil. *[1]*

3 a High temperature increases rate of reaction *[1]*; but high temperature will favour endothermic direction so reduces yield *[1]*; 450 °C (is a moderately high temperature) and is a compromise (between these two competing factors). *[1]*

Changing pressure has no effect on yield (because there are equal numbers of moles on each side of the equation) *[1]*; high pressure would increase rate *[1]*; but would be expensive, because of need for expensive equipment / energy. *[1]*

Catalyst increases rate without need for increased pressure / temperature. *[1]*

b Reaction removes toxic CO *[1]*; produces valuable H_2 *[1]*; although it releases CO_2 which is a greenhouse gas *[1]*; and requires fuel to be burnt which also releases CO_2 *[1]*; overall benefits outweigh the risks. *[1]*

15.3

1 Sun emits visible and ultraviolet, Earth emits infrared *[1]*; wavelength of radiation emitted by Sun is smaller than that emitted by Earth *[1]*; frequency of radiation emitted by Sun is greater than that emitted by Earth. *[1]*

2 Infrared window is the range of wavelengths / frequencies not absorbed by water vapour *[1]*; sulfur hexafluoride absorbs infrared radiation with a wavelength / frequency from within this range. *[1]*

3 Infrared radiation from the Earth is absorbed by bonds in the water molecule *[1]*; some of this energy is re-emitted in all directions, including back to Earth *[1]*; greater evaporation means that the concentration of water vapour will increase *[1]*; a higher concentration of water vapour means that more energy is absorbed and re-emitted in this way *[1]*; temperature of the Earth increases even more. *[1]*

16.1

Go further

There are 16 duplet combinations of the 4 RNA bases.

Summary questions

1 a Consists of 2 chains of nucleotides *[1]*; twisted into a double helix structure *[1]*; sugar and phosphate molecules bond together to form a 'backbone' *[1]*; complementary pairs of bases on the inside of the molecule bond together by hydrogen bonds. *[1]*

b RNA (usually) is single stranded not double stranded *[1]*; RNA has a uracil base instead of a thymine *[1]*; RNA has a ribose sugar instead of a deoxyribose. *[1]*

2 Alanine–Leucine–Glutamic acid–Valine *[1]*

3 mRNA: carries information from the nucleus to the ribosome *[1]*; sequence of codons on mRNA determines sequence of amino acids in protein. *[1]*

tRNA: delivers amino acid to the ribosome *[1]*; anticodon on tRNA recognises / bonds specifically to a codon on mRNA. *[1]*

17.1

1 A molecule with more than one functional group. *[1]*

2 a addition reactions: alkenes, carbonyls. *[1]*

b substitution reactions: haloalkanes, alcohols. *[1]*

c oxidation reactions: primary and secondary alcohols, aldehydes. *[1]*

3 *[1]*

amine and carboxylic acid *[1]*; the amine would act as a proton acceptor so will react with acids, and it could act as a nucleophile to form an amide with acyl chlorides *[1]*; the carboxylic acid would act as a proton donor so will react with bases, and it could be converted to an acyl chloride with $SOCl_2$, and it could undergo esterification with alcohols. *[1]*

17.2

1 a $C_3H_7OH + [O] \rightarrow C_2H_5CHO + H_2O$ *[1]*

b $C_3H_7OH + 2[O] \rightarrow C_2H_5COOH + H_2O$ *[1]*

c $C_3H_7OH \rightarrow C_3H_6 + H_2O$ *[1]*

d $C_3H_7OH + HCl \rightarrow C_3H_7Cl + H_2O$ *[1]*

e $C_3H_7OH + HCOOH \rightarrow C_3H_7OOCH + H_2O$ *[1]*

2 but-1-ene → butan-1-ol *[1]* → butanoic acid *[1]*; conditions for the first step are reaction with H_2O at 300 °C with a catalyst of H_3PO_4 *[1]*; conditions for the second step are refluxing with acidified potassium dichromate. *[1]*

3 haloalkane → alcohol *[1]* → ketone *[1]* → cyanohydrin *[1]*; the ketone is butanone *[1]*; the alcohol is butan-2-ol *[1]*; the haloalkane is 2-bromobutane or 2-chlorobutane *[1]*; the conditions are OH^- ions for making the alcohol from the haloalkane *[1]*, oxidation with acidified potassium dichromate for making the ketone *[1]*, and reaction with HCN for making the cyanohydrin. *[1]*

Answers to practice questions

Chapter 4

1 C *[1]*   **2** D *[1]*   **3** D *[1]*   **4** C *[1]*

5 a Heat loss to the surroundings ACCEPT not done under standard conditions *[1]*

  b i *Top line*: $2Na^+(g) + SO_4^{2-}(g)$ *[1]*

   Bottom line $2Na^+(aq) + SO_4^{2-}(aq)/Na_2SO_4(aq)$ *[1]*

   Top LH arrow: Lattice enthalpy / $\Delta_{LE}H$

   Top RH arrow: sum of enthalpy change of hydration ($\Delta_{hyd}H$) of (positive and negative) ions

   Bottom arrow: enthalpy change of solution

   All 3 arrows correct = *[1]*

  ii sum of enthalpy change of hydration ($\Delta_{hyd}H$) of (positive and negative) ions = $\Delta_{LE}H$ + enthalpy change of solution = $-1944.0 - 2.5 = -1946.5\,kJ\,mol^{-1}$ *[1]*

   Sum of enthalpy changes of hydration = $2 \times -406 + \Delta_{hyd}H\,[SO_4^{2-}]$ *[1]*

   $\Delta_{hyd}H\,[SO_4^{2-}] = -1946.5 + 812$ *[1]* $= -1134.5\,kJ\,mol^{-1}$ *[1]*

6 a Water is polar / has a dipole / partial charges *[1]*

   Water molecules form ion-dipole bonds to ions *[1]*

   Energy released from forming bonds is enough to compensate for energy required to break bonds in lattice OR strength of bonds formed is similar to strength of bonds in lattice *[1]*

  b i $\Delta_{surr}S = -\left(-\dfrac{13100}{298}\right) = +44.0\,J\,K^{-1}\,mol^{-1}$ *[1]*

   $\Delta_{tot}S = -204.8 + 44.0 = -160.8$ *[1]*

   This is negative so process is not feasible *[1]*

  ii (No) $\Delta_{surr}S$ can be made more positive by lowering the temperature *[1]*

   If T is low enough, $\Delta_{surr}S$ will be large enough to compensate for −ve $\Delta_{sys}S$ *[1]*

   T required for this would be (well) below $273\,K$ (so water would not be liquid at this T) *[1]*

Chapter 5

1 C *[1]*   **2** B *[1]*   **3** D *[1]*

4 a i sulfonic acid / sulfonate *[1]*

  ii $-NH_3^+$ (accept any valid protonated amine group e.g. $-NH_2R^+$) *[1]*

  b i Random movement of electrons creates an uneven distribution of charge / instantaneous dipole *[1]*

   This induces a dipole on a neighbouring molecule *[1]* There is an (electrostatic) attraction between the two dipoles / two molecules *[1]*

  ii Long chains / regular arrangement of chains / close packing of chains (any two) *[2]*

  c bond between H atom in O-H group and O or N in fibre *[1]*

lone pair on O pointing to H *[1]*

180° bond angle around H atom *[1]*

δ− and δ+ charges on O-H and N-H bonds *[1]*

5 a i amine *[1]*

  ii amine group has a lone pair of electrons on a small (electronegative) atom OR a δ+ H atom *[1]* can bond to a δ+ H (in the receptor site) OR a lone pair of electrons on a small (electronegative) atom (in the receptor site) *[1]*

  b i contains a chiral centre / C atom bonded to 4 different groups *[1]*

  ii At least one molecule drawn with CH_3, H, and NH_2 groups shown in a 3-dimensional / tetrahedral arrangement around the chiral carbon (at the end of the chain) *[1]*

   two molecules are clearly mirror images *[1]*

Chapter 6

1 B *[1]*   **2** C *[1]*   **3** D *[1]*   **4** D *[1]*

5 $CH_3COOCH_2CH_3$ *[1]*

   δ = 1.2 is $CH_2\underline{CH_3}$ AND 2H on C atom adjacent to CH_3 environment *[1]*

   δ = 2.0 is $\underline{CH_3}CO$, no H atoms on C adjacent to CH_3 environment *[1]*

   δ = 4.2 is $O\underline{CH_2}$, 3 H on C atom adjacent to CH_2 environment *[1]*

   ACCEPT any unambiguous way of identifying environments e.g. labels on structure

6 Electrons are excited to higher energy level *[1]* absorb light energy / electromagnetic radiation *[1]* energy gap between energy levels greater in benzene *[1]* because less extensive delocalised system in benzene *[1]* benzene absorbs in uv region, disperse red absorbs in visible region / complementary wavelengths to red NOT red *[1]*

Chapter 7

1 D *[1]*   **2** C *[1]*   **3** C *[1]*   **4** D *[1]*

5 a $K_c = \dfrac{[C_2H_4][H_2O]}{[C_2H_5OH]}$ *[1]*

  b $mol\,dm^{-3}$ *[1]*

6 a $K_c = \dfrac{[SO_3]^2}{[SO_2]^2[O_2]}$ *[1]*

  b i Increasing pressure moves the position of equilibrium to the right but does not affect the magnitude of K_c. *[1]*

  ii Adding a catalyst does not affect the position of equilibrium or the magnitude of K_c. *[1]*

  iii Increasing the temperature moves the position of equilibrium to the left and decreases the magnitude of K_c. *[1]*

  c $K_c = \dfrac{[SO_3]^2}{[SO_2]^2[O_2]}$

$$= \frac{(1.80 \times 10^{-2}\,\text{mol dm}^{-3})^2}{(7.45 \times 10^{-3}\,\text{mol dm}^{-3})^2 \times (3.62 \times 10^{-3}\,\text{mol dm}^{-3})}$$

$$K_c = \frac{3.24 \times 10^{-4}\,\text{mol}^2\,\text{dm}^{-6}}{(5.55 \times 10^{-5}\,\text{mol}^2\,\text{dm}^{-6}) \times (3.62 \times 10^{-3}\,\text{mol dm}^{-3})}$$

$= 1610$ *[1]* $\text{mol}^{-1}\,\text{dm}^3$ *[1]* (3sf)

7 a $K_{sp} = [Pb^{2+}][S^{2-}]$ *[1]*

b The maximum concentration of S^{2-} ions is
$K_{sp} / [Pb^{2+}]$ *[1]*
$= 1.3 \times 10^{-28}\,\text{mol}^2\,\text{dm}^{-6} / 1.14 \times 10^{-14}\,\text{mol dm}^{-3}$
$= 1.14 \times 10^{-14}\,\text{mol dm}^{-3}$. *[1]*

8 a Retention time is the time taken for a component to emerge from the column in gas–liquid chromatography. *[1]*

b The area under the octane peak will be three times the area under the decane peak. *[1]*

c Mass spectrometry will confirm the relative molecular mass of each component. *[1]*

Chapter 8

1 D *[1]*   **2** B *[1]*   **3** B *[1]*   **4** C *[1]*   **5** D *[1]*

6 A weak acid on its own can dissociate on addition of alkali, but does not have a sufficiently high concentration of [salt] to accept enough protons on addition of acid. *[1]*

7 pH = 4.2 *[1]*. The buffer minimises pH changes *[1]* on addition of small quantities of acid or alkali. Addition of H^+ moves the position of equilibrium to the left *[1]*, and addition of OH^- moves the position of equilibrium to the right *[1]*. This minimises the change in $[H^+]$. *[1]*

Chapter 9

1 D *[1]*   **2** C *[1]*   **3** B *[1]*   **4** D *[1]*   **5** D *[1]*

6 a VO_2^+ is +5 and VO^{2+} is +4. *[2]*

b Reduced because its oxidation state goes from +5 to +4. *[1]*

c $VO_2^+ + 2H^+ + 1e^- \rightleftharpoons VO^{2+} + 1H_2O$ *[1 mark for H_2/H_2O balancing, 1 mark for electron balancing]*

d $E_{cell} = 1.76\,\text{V}$ *[1]*

e $2VO_2^+ + 4H^+ + Zn \rightleftharpoons 2VO^{2+} + 2H_2O + Zn^{2+}$ *[1]*

f Zinc has a more negative $E^{\ominus}$ than all of the vanadium half-cells so it can reduce all the species to V^{2+}. *[1]*

g V^{3+} *[1]*; as $E^{\ominus}$ for Pb/Pb^{2+} is less negative than $E^{\ominus}$ for V^{3+}/V^{2+}, so Pb cannot reduce V^{3+}. *[1]*

7 a $Fe \rightarrow Fe^{2+} + 2e^-$ *[1]*

b $O_2 + 2H_2O + 2e^- \rightarrow 4OH^-$ *[1]*

c $Fe^{2+} + 2OH^- \rightarrow Fe(OH)_2$ *[1]*

d $Zn \rightarrow Zn^{2+} + 2e^-$ *[1]*

Chapter 10

1 D *[1]*   **2** D *[1]*   **3** B *[1]*   **4** B *[1]*   **5** B *[1]*

6 a The rate would double. *[1]*

b The rate would go up by a factor of eight – doubled due to doubling [M] and quadrupled due to doubling [N] *[1]*.

c $k = \dfrac{\text{rate}}{[M][N]^2}$ *[1]* $= \dfrac{0.025}{(0.1) \times (0.1)^2} = 25$ *[1]*; units are $\text{mol}^{-2}\,\text{dm}^6\,\text{s}^{-1}$ *[1]*

d k increases as temperature increases. *[1]*

e Successive half-lives of [M] are equal *[1]* and successive half-lives of [N] are not equal. *[1]*

7 a First order with respect to P and first order with respect to Q. *[2]*

b Taking tangents at time $t = 0$ / measuring successive half-lives. *[1]*

c Not supported as the slow step contains P only. *[1]*

Chapter 12

1 D *[1]*   **2** A *[1]*   **3** A *[1]*   **4** B *[1]*   **5** B *[1]* **6**

a model A has 6 delocalised electrons *[1]*
model B has 3 C–C pi bonds / alternating C–C and C=C bonds *[1]*

b *Level of response mark scheme:*

Evidence from bond lengths (from X-ray diffraction):

bond lengths equal
bond lengths intermediate between C–C and C=C
fits A
does not fit B (because C–C longer than C=C)

Evidence from pattern of chemical reactions

takes part in (mostly) substitution reactions
fits A as substitution maintains delocalised structure does not fit B as C=C would take part in addition reactions

Evidence from thermochemical data

enthalpy of hydrogenation is less negative than predicted / not 3 x value for cyclohexene
fits A as bonding in benzene is stronger than 3 x C=C / delocalisation increases stability
does not fit B as 3 separate C=C bonds / no extra stability

[5–6 marks]

Discusses 3 types of evidence
All evidence described in detail
Clearly describes how evidence links to bonding

[3–4 marks]

Discusses 2 types of evidence
Most evidence described in detail OR
discusses 3 types of evidence but in less detail
Some success in linking evidence to bonding

[1–2 marks]

Discusses 1 type of evidence
Describes this in some detail OR
Discusses 2 types of evidence but in little detail
Limited success in linking evidence to bonding

7 a i conc nitric and conc sulfuric acid *[1]*

ii Sn and HCl *[1]*

b i $C_6H_5–N^+\equiv N$ *[1]*   **ii** C_6H_5OH *[1]*

c Reaction 1: below 55°C *[1]* to avoid multiple nitrations. *[1]*

Reaction 2: below 5°C *[1]* to slow down decomposition of diazonium compound. *[1]*

Chapter 13

1 D *[1]*   **2** B *[1]*   **3** C *[1]*   **4** A *[1]*

5 a $C_5H_{10}O_2$ *[1]*

b $2\ C_4H_9COOH + Na_2CO_3 \rightarrow 2\ C_4H_9COONa + CO_2 + H_2O$ *[1]*; carbon dioxide produces bubbles and turns limewater cloudy. *[1]*

c $C_4H_9COOH + NH_3 \rightarrow C_4H_9CONH_2 + H_2O$ *[1]*; primary amide. *[1]*

6 a $H_2NCH_2CONHCH(CH_3)CONHCH(CH_2OH)COOH$ *[1]*

b If acid hydrolysis is used, amine salts of the amino acids are produced, but if alkaline hydrolysis is used, carboxylate salts are produced. *[1]*

c Only two of the amino acids have a chiral centre *[1]*; as they have four different groups around a carbon atom *[1]*; gly does not display optical isomerism as it does not have a carbon with four different groups. *[1]*

d A spot of the product mixture is placed on the paper *[1]*; the paper is placed in a solvent, and the solvent rises to the top of the paper, separating the components *[1]*; R_f values are calculated and compared to reference data, or use spots of the three amino acids and compare heights. *[1]*

7 a 2-heptanone is a ketone and benzaldehyde is an aldehyde. *[1]*

b $CH_3COCH_2CH_2CH_2CH_2CH_3$ *[1]*

c Only benzaldehyde will react with Fehling's solution *[1]*; a red precipitate will be formed. *[1]*

d Only benzaldehyde will react with Tollens' reagent *[1]*; a silver mirror will be formed. *[1]*

e Both will react with HCN *[1]*. $CH_3C(CN)(OH)CH_2CH_2CH_2CH_2CH_3$ *[1]* and $C_6H_5CH(OH)(CN)$ *[1]*

Chapter 14

1 A *[1]*   **2** D *[1]*   **3** B *[1]*   **4** C *[1]*   **5** C *[1]*

6 a $(4 \times 30) \times 100 / ((4 \times 30) + (6 \times 18)) = 52.6\%$ *[1]* low atom economy means that large mass of waste is formed *[1]*

b i co-product = H_2O *[1]* additional product formed as a result of the intended reaction *[1]*

ii by-product = NO_2 *[1]* product formed as a result of an unintended reaction *[1]*

c i higher temperature is not used as reaction is exothermic *[1]* higher temperature favours endothermic direction AND reduces yield *[1]*

ii higher pressure would decrease yield *[1]* equilibrium would favour LHS AND fewer moles of gas *[1]* higher pressure increases costs *[1]* more energy needed to compress gas OR expense of thick-walled reaction vessels *[1]* lower pressure decreases rate *[1]* 5 atm is a

compromise / provides acceptable rate without decreasing yield significantly (or reverse argument) *[1]*

Chapter 15

1 D *[1]*   **2** C *[1]*   **3** C *[1]*   **4** D *[1]*   **5** A *[1]*

6 Molecules of carbon dioxide absorb infrared radiation. *[1]* Infrared radiation is emitted by the Earth's surface. *[1]* Bonds vibrate with greater energy. *[1]* Some of this energy is radiated back towards Earth. *[1]* Some of the energy is passed on to other molecules by collisions. *[1]* Higher concentration of CO_2 means that a greater fraction / proportion of radiation is absorbed. *[1]*

Chapter 16

1 D *[1]*   **2** A *[1]*   **3** B *[1]*   **4** D *[1]*   **5** B *[1]*   **6** A *[1]*

7 a The hydrogen bonds between the (complementary) base pairs break AND the two strands separate. *[1]*
New RNA nucleotides bond to the exposed bases. *[1]*
The newly attached nucleotides are joined together, creating a new strand of RNA. *[1]*

b i mRNA / messenger RNA *[1]*

ii codon (IGNORE any reference to amino acid encoded) *[1]*

iii anti-codon *[1]*

iv peptide (ACCEPT amide) *[1]*

c Codons consist of 3 bases. *[1]*

There are more than 20 combinations / 64 combinations of bases in codons *[1]* ACCEPT any number >20. *[1]*

Chapter 17

1 D *[1]*   **2** A *[1]*   **3** A *[1]*   **4** A *[1]*   **5** A *[1]*

6 Ketone, amine, alcohol, ether *[4]*; ketones can be reduced to alcohols or react with HCN *[1]*; amines will react with acids or can form amides *[1]*; alcohols can react with carboxylic acids to form esters. *[1]*

7 **Step 1:** CH_3CH_2CHO to $CH_3CH_2CH_2OH$; reduction with $NaBH_4$. *[1]* **Step 2:** $CH_3CH_2CH_2OH$ to $CH_3CH_2CH_2Br$; nucleophilic substitution with HBr. *[1]* **Step 3:** $CH_3CH_2CH_2Br$ to $CH_3CH_2CH_2CN$; nucleophilic substitution with HCN. *[1]* An extra step after Step 1 could involve dehydrating the alcohol to form an alkene, then reacting the alkene with HBr to form $CH_3CH_2CH_2Br$.

8 **Step 1:** C_6H_6 to $C_6H_5NO_2$; nitration with concentrated nitric acid and concentrated sulfuric acid at <55°C. *[1]*

Step 2: $C_6H_5NO_2$ to $C_6H_5NH_2$; reduction with tin and concentrated hydrochloric acid. *[1]*

Synoptic questions

1 Aluminium chloride is manufactured from aluminium metal. It has several uses, for example as a catalyst in several important reactions of benzene. Aluminium can be formed from the reaction of aluminium with hydrogen chloride at about 700°C.

Reaction A: $2\,Al + 6\,HCl \rightarrow 2\,AlCl_3 + 3\,H_2$

a A small-scale pilot study of this process was carried out.

  i The yield in this pilot study was found to be 95%. Calculate the mass of aluminium needed to produce 10 kg of aluminium chloride. *(2 marks)*

  ii Calculate the atom economy of reaction A to 3 significant figures. *(1 mark)*

b A second reaction that could be used to form aluminium chloride is shown below:

Reaction B: $2Al + 3Cl_2 \xrightarrow{700°C} 2AlCl_3$

Use the principles of green chemistry to compare the environmental impact of reactions A and B. *(2 marks)*

c Aluminium chloride is a solid ionic compound at temperatures below 180°C. However, above this temperature the compound is converted into a liquid **covalent** compound with the formula Al_2Cl_6.

  i What can you predict about the electrical conductivity of aluminium chloride at room temperature, and at temperatures above 180°C? Explain your answer. *(2 marks)*

  ii The bonding of the covalent molecule Al_2Cl_6 can be represented by the diagram shown below:

Draw a dot-and-cross diagram to show the bonding in this compound. *(2 marks)*

d Aluminium chloride is used as a catalyst in a range of reactions, known as Friedel–Crafts reactions, which are important in synthesis.

  i Which reaction is a Friedel–Crafts reaction? Explain the importance of these reactions in synthesis. *(1 mark)*

  ii Give the structure of the organic reactant used in reaction 2. *(1 mark)*

  iii Identify the **types** and **mechanisms** of the reactions that occur in reaction 1 AND reaction 2. *(2 marks)*

  iv Name the **type** of product formed in reaction 2. *(1 mark)*

  v State the reagent and conditions required for reaction 3. *(1 mark)*

2 Sodium carbonate exists as a hydrated salt, $Na_2CO_3.xH_2O$.

a Some students carry out an experiment to find calculate the value of x. They heat some hydrated sodium carbonate in a crucible and record the following results:

	Mass / g
1. Mass of crucible	10.36 g
2. Mass of crucible + hydrated salt	14.77 g
3. Mass of crucible + salt after 5 minutes heating	11.99 g
4. Mass of crucible + salt after a further 2 minutes heating	11.99 g

  i What was the purpose of taking reading 4 (mass of crucible + salt after a further 2 minutes heating)? *(1 mark)*

  ii Calculate the value of x in the formula $Na_2CO_3.xH_2O$. *(4 marks)*

b Magnesium carbonate also exists in hydrated forms. However it is not possible to use the same method to find the formula of magnesium carbonate as anhydrous magnesium carbonate undergoes thermal decomposition at a low temperature.

  i Write an equation for the thermal decomposition of anhydrous magnesium carbonate. *(1 mark)*

  ii Explain why anhydrous magnesium carbonate decomposes at a lower temperature than anhydrous sodium carbonate. *(2 marks)*

  iii A titration method is used to find the formula of **hydrated** magnesium carbonate.

1.69 g of a sample of hydrated magnesium carbonate was dissolved in water and the volume made up to 250 cm³.

25.0 cm³ of this solution was titrated against 0.100 mol dm⁻³ HCl. The average titre was 24.50 cm³.

Calculate the value of x. *(4 marks)*

c Carbonate rocks can be heated to high temperatures in the interior of the Earth, releasing carbon dioxide. Some of this carbon dioxide reaches the atmosphere and acts as a greenhouse gas.

Describe how carbon dioxide acts as a greenhouse gas. *(6 marks)*

d Carbon dioxide from the atmosphere and carbonate ions from rocks are involved in a series of processes involving water that form a weakly acidic solution:

Equation 1: $CO_2(g) + aq \rightleftharpoons CO_2(aq)$

Equation 2: $CO_2(aq) + H_2O(l) \rightleftharpoons H^+(aq) + HCO_3^-(aq)$

Equation 3: $HCO_3^-(aq) \rightleftharpoons H^+(aq) + CO_3^{2-}(aq)$

i Explain how a decrease in the pH of the ocean can result in the release of carbon dioxide into the atmosphere. *(4 marks)*

ii A mixture of sodium carbonate and sodium hydrogencarbonate can also be used as a buffer solution.

Calculate the pH of a buffer solution formed by mixing equal volumes of $0.20 \, mol \, dm^{-3}$ $NaHCO_3$ and $0.40 \, mol \, dm^{-3}$ Na_2CO_3.

$Ka [HCO_3^-] = 4.8 \times 10^{-11} \, mol \, dm^{-3}$. *(2 marks)*

e Carbon dioxide is also converted very rapidly into hydrogen carbonate ions by the enzyme carbonic anhydrase.

This catalyses the reaction shown in Equation 2, above.

The optimum temperature of the enzyme is 37°C and the optimum pH is 7.5.

i Sketch the likely pattern in activity when the temperature is altered from 20°C to 65°C. *(2 marks)*

ii Explain why changing the pH from 7.5 to 6.5 produces a significant drop in enzyme activity. *(2 marks)*

3 Propanone reacts with iodine in the presence of an acid catalyst

$CH_3COCH_3 + I_2 \rightarrow CH_3COCH_2I + HI$

a An experiment was carried out in which the concentration of iodine was measured at regular time intervals. The results were plotted on a graph

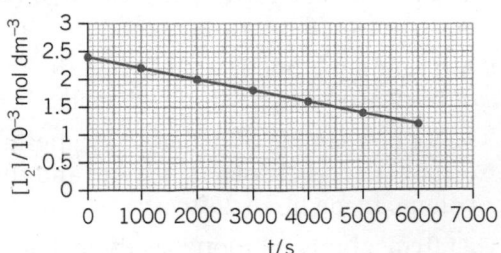

i Describe how these data could have been obtained experimentally. *(4 marks)*

ii State how the rate of reaction changes during the experiment, and explain how this pattern enables you to deduce that the order with respect to iodine is zero. *(2 marks)*

b Further experiments were carried out in which the rate was measured with different concentrations of reactant.

Experiment	$[H+] /$ $mol \, dm^{-3}$	$[CH_3COCH_3] /$ $mol \, dm^{-3}$	Rate $/ mol \, dm^{-3} \, s^{-1}$
1	1.00	0.80	5.6×10^{-4}
2	2.00	0.80	1.2×10^{-3}
3	2.00	0.40	5.6×10^{-4}
4	1.50	0.40	4.2×10^{-4}

i The overall rate equation is:

rate $= k[H^+][CH_3COCH_3]$.

Explain how the orders with respect to H^+ and propanone can be deduced from these data. *(2 marks)*

ii Calculate a value for k for this reaction, including its units. *(2 marks)*

c The following mechanism has been proposed:

step 1: fast

step 1: slow

step 3: fast

i Complete the mechanism by adding the missing curly arrows and the formula of product X. *(4 marks)*

ii Explain how this mechanism is consistent with the rate equation shown in **b**. *(3 marks)*

d During the reaction, a solution of hydrogen iodide is formed.

i A solution of HI is known as hydroiodic acid, which is a strong acid.

Calculate the pH of $0.200 \, mol \, dm^{-3}$ hydroiodic acid. *(1 mark)*

ii A student tries to form a solution of hydroiodic acid by passing hydrogen iodide gas into water.

To form hydrogen iodide, he adds concentrated sulfuric acid to solid sodium iodide. Steamy fumes of hydrogen iodide are formed but also a purple vapour.

Identify the purple vapour and state **one** other gaseous product formed in the reaction.

(2 marks)

iii Describe a better way of preparing hydrogen iodide from sodium iodide. *(1 mark)*

e If hydrogen iodide gas is heated in a sealed vessel an equilibrium is set up $2HI \rightleftharpoons H_2 + I_2$

$0.0200\,mol$ of hydrogen iodide are heated to $600\,K$ in a vessel of volume $1\,dm^3$. At equilibrium, $0.0090\,mol$ of I_2 is formed.

i Use this information to calculate a value of K_c. Give your answer to an appropriate number of significant figures. *(3 marks)*

ii At $700\,K$, K_c has a value of 50. Explain what you can deduce about the sign of ΔH for this process. *(2 marks)*

f Hydrogen fluoride has properties that differ significantly from those of hydrogen iodide. For example it has a much higher boiling point, it is a weak acid when dissolved in water and HF does not decompose even when heated to very high temperature.

A student tries to estimate the enthalpy change for the formation of hydrogen fluoride. He uses the following bond enthalpies data:

$E(H–H) = +436\,kJ\,mol^{-1}$

$E(H–F) = +568\,kJ\,mol^{-1}$

$E(F–F) = +158\,kJ\,mol^{-1}$

i Calculate a value for the $\Delta_f H$ [HF]. *(2 marks)*

ii The student knows estimates of the enthalpy change using bond enthalpies are often different to the true values. Discuss whether the value you estimated in **i** is likely to be significantly different to the true value.

(2 marks)

iii A solution of hydrogen fluoride in water is a weak acid.

K_a [HF] $= 5.6 \times 10^{-4}\,mol\,dm^{-3}$

Calculate a value for the pH of $0.100\,mol\,dm^{-3}$ hydrofluoric acid. Comment on the validity of any assumptions that you make. *(4 marks)*

iv Use ideas about the structure and bonding in hydrogen fluoride to discuss reasons for the differences in properties between hydrogen fluoride and other hydrogen halides *(6 marks)*

---

4 1,2-dichloroethene, CHCl=CHCl, is the starting material for the production of a range of organic molecules, including polymers.

a i Explain why this molecule can exist as a pair of stereoisomers. *(2 marks)*

ii Bonds in organic molecules are described as π (pi) or σ (sigma) bonds. State the number of each type of bond present in this molecule:

Number of π bonds: _____ Number of σ bonds: _____ *(1 mark)*

b The table below gives details of three reactions, A–D of 1,2-dichloroethene.

Complete the table with the missing reagents and structures.

Reaction	Reagents	Structure of product
A	**i**	CH_2ClCH_2Cl
B	Bromine liquid	**ii** ....
C	**iii**	$CHClOHCH_2Cl$
D	Mixture of aqueous bromine and sodium chloride solution	**iv** "............ or ...." two possible products

(4 marks)

c i Give the full structural formula of the repeat unit of the polymer formed from 1,2-dichloroethene. *(1 mark)*

ii Explain what type of polymerisation has occurred in the formation of this polymer.

(1 mark)

d A reaction can also occur with HF to form a fluorinated compound, $CHClFCH_2Cl$. This compound has been used as a solvent. It can also exist as a pair of stereoisomers.

i Give the systematic name of this molecule. *(1 mark)*

ii Explain why this molecule can exist as a pair of stereoisomers. You should include 3-dimensional structures to illustrate your answer. *(3 marks)*

e The fluorinated compound has been linked to the destruction of ozone in the stratosphere.

i Explain why the presence of ozone in the stratosphere is significant for human health.

(2 marks)

The destruction of ozone is thought to occur because of the formation of Cl radicals. These react with ozone in a catalytic cycle.

ii Complete these equations to show the processes that destroy ozone in the stratosphere:

$Cl + O_3 \rightarrow$ _____ + _____

_____ + _____ $\rightarrow Cl + O_2$ *(2 marks)*

iii Name the type of radical reaction that is occurring in this catalytic cycle. Give a reason for your answer. *(2 marks)*

f The highest frequency radiation that reaches the troposphere is described as UVA. This has a wavelength range of 315 nm to 400 nm ($1\,nm = 10^{-9}\,m$).

The bond energy of a C–Cl bond in $CHClFCH_2Cl$ is estimated to be $340\,kJ\,mol^{-1}$.

Discuss whether UVA radiation could cause the production of Cl radicals from $CHClFCH_2Cl$ molecules in the troposphere. *(5 marks)*

5 Carboxylic acids and esters are found in many natural products. For example, lactic acid, a white water-soluble solid, and the ester ethyl lactate, a colourless liquid with a boiling point of 155°C, are both found in wine.

lactic acid   ethyl lactate

a i Give the systematic name of lactic acid. *(1 mark)*

ii State the reagents and conditions used to form ethyl lactate from lactic acid. *(2 marks)*

b A student attempts to separate and purify a sample of ethyl lactate from the reaction mixture.

i Suggest how the liquid ester would be separated from the reaction mixture. *(1 mark)*

ii Give the name of a substance used to remove water from the sample of the ethyl lactate. *(1 mark)*

c The purified product was analysed using infrared spectroscopy, along with a sample of lactic acid.

Explain how infrared spectroscopy could be used to show that lactic acid has been successfully converted into ethyl lactate. *(3 marks)*

d When lactic acid is heated with a phosphoric acid catalyst it is converted into a new liquid compound, X.

The empirical formula is $C_6H_{10}O_5$ and the molecular ion of the mass spectrum occurs at an M/z value of 162.

The infrared spectrum has a broad peak between 3200 and 3600 cm^{-1} and a peak at 1790 cm^{-1}.

The ^{13}C nmr spectrum has 3 peaks at $\delta = 220$, $\delta = 85$, and $\delta = 35$.

The proton NMR spectrum has 3 peaks. The details of these are shown below:

Chemical shift	Splitting	Relative number of H in the environment
1.8	doublet	3
4.2	quartet	1
11.5	singlet	1

Use the information provided to deduce the structure of the molecule X, explaining how the structure links to the information. *(6 marks)*

e Lactic acid can also form a polymer, poly(lactic acid).

i Draw out the repeat unit of this polymer and explain the type of polymerisation that has occurred. *(2 marks)*

ii Suggest why poly(lactic acid) is regarded as a 'greener' product than polymers such as nylon. *(1 mark)*

f Another carboxylic acid found in living organisms is ethanedioic acid, $H_2C_2O_4$, found in the leaves of plants such as rhubarb. Salts of ethanedioic acid, such as calcium ethanedioate, are also found naturally, and the ethanedioate ion is able to act as a bidentate ligand in several metal complexes.

i Write a balanced equation for the formation of calcium ethanedioate from the reaction between calcium hydroxide and ethanedioic acid. *(1 mark)*

ii Draw out the full structural formula of the ethanedioate ion and explain why it can act as a bidentate ligand. *(3 marks)*

iii Ethanedioate forms an 6-coordination complex ion with Fe^{3+} ions. Deduce the formula of this complex ion. *(2 marks)*

iv Draw out a 3-dimensional structure of this complex ion. *(2 marks)*

g Ethanedioc acid can act as a reducing agent. For example, it is able to reduce a solution of acidified manganate ions.

Electrode potential data for the two half-equations involved in this process are given below:

$$MnO_4^-(aq) + 8H^+(aq) + 5e^- \rightarrow Mn^{2+}(aq) + 4H_2O(l)$$
$$E^\emptyset = +1.52V$$

$$2CO_2(g) + 2H+ (aq) + 2e^- \rightarrow H_2C_2O_4(aq)$$
$$E^\emptyset = -0.49V$$

i Use the electrode potential data to describe why ethandioic acid is able to reduce a solution of acidified manganate ions. *(3 marks)*

ii Describe what you would see during this reaction. *(2 marks)*

iii An impure sample of ethanedioic acid with a mass of 0.98 g was dissolved to form 250 cm^3 of solution.

25.0 cm^3 of the solution was titrated with 0.0200 $mol\,dm^{-3}$ acidified MnO_4^- solution.

The average titre was 19.10 cm^3.

Calculate the % purity of the ethanedioic acid. Give your answer to an appropriate number of significant figures. *(5 marks)*

Answer to synoptic questions

1 a i $\text{mol } AlCl_3 = \dfrac{10,000}{130.5} = 76.6$ [1] mol Al required

$$= 76.6 \times \dfrac{100}{95} \text{ AND}$$

mass $= 80.66 \times 24 = 1.935\,kg$ [1]

ii $\dfrac{(2 \times 130.5) \times 100}{(2 \times 130.5 + 2)} = 99.2\%$ [1]

b ANY two from: similar energy requirements / more waste produced in B / similar raw materials used (aluminium ore and sea water / rocksalt) / chlorine is more hazardous than hydrochloric acid [2]

c i does not conduct below 180°C AND ions are not free to move [1] does not conduct above 180°C AND no ions / charged particles present [1]

ii

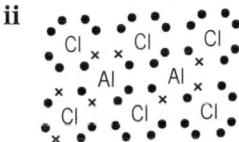

$2 \times$ dative covalent bond [1] all other details [1]

d i reaction 1 AND forms C–C bonds [1]

ii CH_3CH_2COCl [1] (ACCEPT full or skeletal formula)

iii electrophilic substitution [1] nucleophilic addition [1]

iv cyanohydrin [1]

v any named strong acid AND reflux [1]

2 a i to ensure that all water had been removed [1]

ii moles of anhydrous salt $= \dfrac{1.63}{106} =$

$0.01538\,mol$ [1] moles of water $= \dfrac{2.78}{18} =$

0.1544 [1] $x = \dfrac{0.1544}{0.01538} = 10$ [1] ecf from

incorrect calculations of moles above [1]

b i $MgCO_3(s) \rightarrow MgO(s) + CO_2(g)$ [1]

ii magnesium has a higher charge density / smaller radius [1] has a greater ability to polarise the carbonate ion [1]

iii moles HCl $= 2.45 \times 10^{-3}$ [1] mol $MgCO_3$

in $25.0\,cm^3 = \dfrac{2.45 \times 10^{-3}}{2}$ AND

moles $MgCO_3$ in $250\,cm^3 = \dfrac{2.45 \times 10^{-2}}{2} =$

$0.01225\,mol$ [1] M_r of magnesium carbonate

$= \dfrac{1.69}{0.01225} = 137.8$ [1] total M_r of water

molecules $= 137.8 - 84.3 = 53.5$ AND

$x = \dfrac{53.5}{18} = 3.0$ (to 2 s.f.) [1]

c absorbs infrared radiation [1] emitted by Earth [1] bonds vibrate with greater energy [1] reradiates this energy AND some is radiated back to Earth [1] passes on energy to other molecules by

collision [1] temperature of atmosphere increases [1]

d i [H+] increases [1] [CO_2] in equation 2 increases [1] to keep K_c constant [1] [CO_2(aq)] in equation 1 also increases, resulting in increase in [CO_2(g)] [1]

ii [H+] $= K_a \times$ [HA] / [A-] $= 4.8 \times 10^{-11} \times \dfrac{0.20}{0.40} =$

$2.4 \times 10^{-11}\,mol\,dm^{-3}$ [1] pH $= -\log 2.4 \times 10^{-11} = 10.6$ [1]

e i graph showing maximum activity at 37°C AND rapid fall above 37°C [1] exponential increase between 20 and 37 [1]

ii pattern of charges in enzyme change / NH_2 becomes NH_3^+ / COO- becomes COOH [1] affects ability of active site to bind to substrate OR alters shape of active so substrate cannot fit in [1]

3 a i Extract small samples at regular time intervals [1] quench by cooling / adding a base [1] titrate against sodium thiosulfate [1] concentration can be calculated from the value of the titre [1]

OR use a colorimeter with appropriate filter AND zero with distilled water [1] measure absorbance at regular time intervals [1] measure absorbance of known concentrations of iodine [1] plot graph of absorbance vs conc AND read off concentration [1]

ii Rate of reaction remains constant [1] so rate is not affected by changes in [I_2] (hence order is zero w.r.t. I_2) [1]

b i Comparing expt 1 + 2 AND rate doubles as [H+] doubles [1] Comparing expts 2 + 3 AND rate halves as [propanone] halves [1]

ii $k = $ rate / [H+][propanone] $= \dfrac{5.6 \times 10^{-4}}{1 \times 0.8} =$

7.0×10^{-4} [1] units $= dm^3\,mol^{-1}\,s^{-1}$ [1]

c i 6 correct curly arrows $= $ [4], 4 or 5 $= $ [3], 3 $= $ [2], 2 $= $ [1] X $= $ H+ [1]

H+

O
‖
H_3C C CH_3

step 1: fast

OH+
‖
H_3C C C H
H H

step 2: fast

step 3: fast

ii step 2 is the rate-determining step [1] propanone and H^+ appear before the rate-determining step so will be 1st order [1] I_2 appears after the rate-determining step so will be zero order [1]

d i pH = $-\log 0.2 = 0.7$ [1]

ii iodine [1] hydrogen sulfide [1]

iii add phosphoric acid (to solid sodium iodide) [1]

e i equilibrium concentration of HI = $0.020 - 0.090 = 0.002$ [1] $K_c = \dfrac{0.009^2}{0.002^2}$ ecf from incorrect calculation of [HI] or if [HI] taken as 0.020 [1] $K_c = 20$ ecf [1]

ii ∆H is positive AND K_c has increased [1] K_c increases with T for endothermic reactions OR increasing temperature favours the exothermic direction [1]

f i (reaction is $\frac{1}{2}H_2 + \frac{1}{2}F_2 \rightarrow HF$) Bonds formed = 568 kJ mol^{-1}, bonds broken = $\frac{1}{2} \times 436 + \frac{1}{2} \times 158 = 297$ kJ mol^{-1} [1] ∆H = broken − formed = −271 kJ mol^{-1} [1]

ii (unlikely to be different) average bond enthalpies are not used (these are the only compounds that contain these bonds) [1] all molecules are in the gaseous state [1]

iii $[H^+] = \sqrt{(5.6 \times 10^{-4} \times 0.050)} = 5.29 \times 10^{-3}$ mol dm^{-3} [1] pH = 2.28 [1] (2 s.f. or more)

Assumptions are that 1. $[HA]_{eqm} = [HA]_{initial}$ AND 2. $[H^+] = [A^-]$ [1]

Assumption 1 is not justified as $[H^+]$ is relatively large in comparison to [HA] 10% of [HA] value [1] IGNORE any comments about assumption 2

iv Points to be made:

- H–F bonds are relatively strong,
- because F atoms are small / bond lengths are short,
- explains stability of H–F at high temperatures,
- and may also be a factor in being a weaker acid,
- H–F bonds are very polar,
- because F is very electronegative,
- hydrogen bonding is possible between H–F molecules,

- explains high boiling point,
- also may be a factor in being a weaker acid

5–6 marks:

Discusses all three properties
Describes three features of structure and bonding
Relates all properties clearly to features of structure and bonding 3–4 marks:

Discusses two properties
Describes two features of structure and bonding
Relates some properties to features of structure and bonding 1–2 marks:

Discusses one property
Describes one features of structure and bonding
Has limited success in relating some properties to features of structure and bonding

4 a i Contains a C=C bond that cannot rotate [1] there are two different groups attached to each C atom (of the double bond) [1]

ii 1 pi bond AND 5 sigma bonds [1]

b i H_2 AND Pt catalyst at RTP OR Ni catalyst at moderately high T and P / 150°C and 5 atm [1]

ii CHClBrCHClBr [1]

iii phosphoric acid at high T and P-300°C and 60 atm OR conc. Sulfuric acid followed by water [1]

iv CHClBrCHCl$_2$ OR CHClBrCHClBr [1]

c i [1]

ii Addition AND no small molecule is lost during polymerisation [1]

d i 1,2-dichlorofluoroethane [1] IGNORE number before fluorine

ii (One of the carbons) has four different groups around it [1] so can exist as non-superimposable mirror images [1]

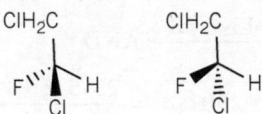

[1] must show correct 3-dimensional geometry

e i absorbs <u>high frequency / high energy</u> UV radiation [1] which would cause skin cancer / eye damage / sunburn [1]

ii ClO + O_2 in 1st equation [1] ClO + O in 2nd equation [1]

iii propagation [1] radicals on both sides of the equation [1]